REWARD

Elementary

Student's Book

Simon Greenall

MACMILLAN
HEINEMANN
English Language Teaching

Map of the book

Lesson	Grammar and functions	Vocabulary	Skills and sounds
20 *A grand tour* Visiting cities in Europe	Past simple (3): irregular verbs *Yes/no* questions and short answers	Verbs connected with tourism Irregular verbs	**Sounds:** pronunciation of some European cities and countries **Listening:** listening for main ideas; listening for specific information **Speaking:** talking about a holiday
Progress check *lessons 16–20*	Revision	Word association Word charts Words formed from other parts of speech	**Sounds:** /ɔː/, /ɜ/ and /ɒ/; silent letter patterns; interested intonation in sentences; syllable stress **Speaking:** playing a game called Twenty Questions
21 *Mystery* An extract from an article about Agatha Christie	Past simple (4): negatives *Wh-* questions	New words from a passage about Agatha Christie	**Reading:** predicting; reading for main ideas; reading for specific information **Sounds:** weak stress of auxiliary verbs **Writing:** writing a short autobiography **Speaking:** talking about your life
22 *Dates* Special occasions and important dates	Past simple (5) Expressions of time	Ordinal numbers Dates, months of the year	**Sounds:** pronunciation of ordinal numbers and dates **Listening:** listening for main ideas; understanding text organisation **Speaking:** talking about your answers to a quiz
23 *What's she wearing?* Clothes and fashion	Describing people (2) Present continuous or present simple	Items of clothing and accessories Colours Actions of the face, hand and body	**Listening:** listening for specific information **Reading:** reading and answering a questionnaire **Speaking:** discussing your answers to the questionnaire
24 *I'm going to save money* Talking about resolutions and future plans	*Going to* *Because* and *so*	New verbs from this lesson	**Reading:** predicting; reading for main ideas **Listening:** predicting; listening for main ideas **Writing:** writing resolutions; joining sentences with *because* and *so*
25 *Eating out* Eating in different kinds of restaurants	*Would like* Talking about prices	Food items	**Listening:** predicting; listening for specific information **Sounds:** pronunciation of *like* and *'d like* **Reading:** reading for specific information **Writing:** writing a paragraph about eating out in your country
Progess check *lessons 21–25*	Revision	Techniques for dealing with difficult words Word puzzle	**Sounds:** words with the same pronunciation but different spelling; /ʊ/ and /uː/; syllable stress; polite intonation **Writing:** preparing a quiz **Speaking:** asking and answering quiz questions
26 *Can I help you?* Shopping	Reflexive pronouns Saying what you want to buy Giving opinions Making decisions	Items of shopping	**Listening:** listening for specific information; listening for main ideas **Sounds:** polite and friendly intonation **Speaking:** talking about shopping habits
27 *Whose bag is this?* Describing objects and giving information	*Whose* Possessive pronouns Describing objects	Adjectives to describe objects Materials	**Speaking:** describing objects; acting out conversations in a Lost Property office **Listening:** listening for main ideas; listening for specific information
28 *What's the matter?* Minor illnesses Healthcare in Britain	Asking and saying how you feel Sympathising *Should, shouldn't*	Adjectives to describe how you feel Nouns for illnesses Parts of the body	**Listening:** listening for main ideas **Reading:** reading for specific information; dealing with unfamiliar words **Writing:** writing an advice leaflet about healthcare for visitors in your country
29 *Country factfile* Facts about Thailand, United Kingdom and Sweden	Making comparisons (1): comparative and superlative forms of short adjectives	Adjectives to describe countries Measurements	**Sounds:** syllable stress; pronunciation of measurements **Reading:** reading for specific information **Listening:** listening for specific information **Writing:** writing a factfile for your country

Lesson	Grammar and functions	Vocabulary	Skills and sounds
30 *Olympic spirit* Olympic records	Making comparisons (2): comparative and superlative forms of longer adjectives	Sports Adjectives to describe sports	**Listening:** listening for main ideas **Speaking:** talking about Olympic sports **Reading:** understanding text organisation
Progress check lessons 26–30	Revision	Adjectives and opposites Word Zigzag	**Sounds:** words which rhyme; /ei/ and /aɪ/; stress patterns **Speaking:** conversation building for sentences **Listening:** listening for main ideas
31 *When in Rome, do as the Romans do* Customs and rules in different countries	*Needn't, can, must, mustn't*	New words from this lesson	**Reading:** reading for main ideas **Speaking:** talking about rules and customs in different countries **Listening:** listening for main ideas; listening for specific information **Sounds:** pronunciation of *must* and *mustn't* **Writing:** writing advice and rules for visitors to your country
32 *Have you ever been to London?* Travel experiences	Present perfect (1): talking about experiences	New words from this lesson	**Reading:** reading for main ideas **Sounds:** strong and weak forms of *have* and *haven't* **Speaking:** talking about things to do in your town; talking about experiences **Writing:** writing a postcard
33 *What's happened?* Talking about good and bad luck	Present perfect (2): talking about recent events *Just* and *yet*	Adverbs and their opposites	**Listening:** listening for specific information **Speaking:** talking about things you have and haven't done; acting out conversations about good and bad luck
34 *Planning a perfect day* Favourite outings	Imperatives Infinitive of purpose	Words connected with outings	**Speaking:** talking about a perfect day **Reading:** reading for main ideas **Writing:** writing advice for planning the perfect day out
35 *She sings well* Schooldays	Adverbs	Adverbs and their opposites	**Sounds:** identifying attitude and mood **Reading:** reading for main ideas; reading for specific information **Listening:** listening for main ideas **Speaking:** talking about achievement at school; performing an action in the style of different adverbs
Progress check lessons 31–35	Revision	Collocation	**Sounds:** words with the same vowel sound; /əʊ/ and /ɔɪ/; word stress and a change of meaning **Reading:** reading for specific information **Writing:** focusing on unnecessary words
36 *I'll go by train* Travel by train and plane	Future simple (1): (*will*) for decisions	Features of an airport and station	**Listening:** listening for specific information **Reading:** reading for specific information **Speaking:** acting out a role-play in a travel agent's
37 *What will it be like in the future?* Talking about the future	Future simple (2): (*will*) for predictions	Nouns and adjectives for the weather	**Listening:** listening for specific information **Reading:** predicting; reading for specific information
38 *Hamlet was written by Shakespeare* World facts	Active and passive	Verbs used for passive	**Speaking:** talking about true and false sentences **Reading:** reading and answering a quiz **Listening:** listening for specific information **Writing:** writing a quiz about your country
39 *She said it wasn't far* Staying in a youth hostel	Reported speech: statements	Items connected with travel	**Reading:** reading for main ideas **Listening:** listening for specific information **Writing:** writing a letter of complaint
40 *Dear Jan ... Love Ruth* A short story by Nick McIver	Tense review	New words from this lesson	**Reading:** predicting; reading for main ideas **Listening:** listening for specific information **Writing:** writing a different ending to the story
Progress check lessons 36–40	Revision	Prepositions Word association	**Sounds:** /ɔ/ and /ɔɪ/; syllable stress **Speaking:** playing *Reward Snakes and Ladders*

1 | *What's your name?*

Present simple (1): *to be*; asking and saying who people are and where they're from

READING AND LISTENING

1 Match the sentences and the people in the photos.

a Hello, I'm Lah. I'm from Thailand.
b Hello, I'm Renato. I'm from Brazil.
c Hello, I'm Sally. I'm from the United States of America.
d Hello, I'm Mehmet. I'm from Istanbul.

2 🔲 Read and listen.

M Hello, I'm Marco. What's your name?
E Hello, Marco. My name's Enrique.
M Where are you from, Enrique?
E I'm from Madrid.

3 Put the sentences in the right order to make a conversation.

☐ I'm George. Where are you from, Marie?
☐ I'm from Paris. And you?
☐ I'm from Athens.
☐ Hello. What's your name?
☐ Hello, my name's Marie. What's your name?

4 Work in pairs and check your answers to 3.
🔲 Now listen and check.

GRAMMAR AND FUNCTIONS

> Present simple (1): *to be*
> **I'm** *Lah.* (= *I am Lah.*)
> **My name's** *Enrique.* (= *My name is Enrique.*)
> **What's** *your name?* (= *What is your name?*)
>
> Asking and saying who people are
> *What's your name?*   *My name's Anna.*
>   *I'm Anna.*
> Asking and saying where people are from
> *Where are you from?*   *I'm from Italy.*

1 Put the words in the right order and write sentences.

   1   I'm hello Manuel.   4   London from I'm.
   2   you are from where?   5   name's hello my Toni.
   3   your name what's?   6   Hannah name's my.

2 Match the sentences in 1.

   Hello, I'm Manuel.   *Hello, my name's Toni.*

3 Complete the sentences.

   1   What _____ your name?   4   My name _____ Peter.
   2   Where are you _____?   5   Hello, _____'m Johann.
   3   I _____ from China.   6   _____ are you from?

   1   What is your name?

VOCABULARY AND SOUNDS

1 Look at the words in the vocabulary box. Put them in two columns: *cities* and *countries.*

> Britain   Bangkok   Tokyo   Budapest   Japan   Canada
> London   Mexico   Thailand   Hungary   Australia   Brazil

   cities: Bangkok   *countries: Britain*

2 🔊 Listen and repeat these words.

   ☐ ☐   ☐ ☐☐   ☐   ☐   ☐☐ ☐   ☐ ☐
   Britain   Tokyo   Bangkok   Budapest   Japan
   Canada

How many syllables does each word have?

Britain, two   Tokyo, three ...

3 Look at these words and underline the stressed syllable.

London   Mexico   Thailand   Hungary
Australia   Brazil

🔊 Listen and check.

4 What's your country's name in English?

5 🔊 Listen and repeat.

name   your name   What's your name?
name   My name's Pat.
from   where   Where are you from?
from   I'm from Britain.

SPEAKING

1 Ask and say who you are and where you're from.
Use the conversation in *Reading and listening* activity 2 and 3 to help you.

2 Think of a famous person and where they come from. Go round asking and saying who you are and where you come from.

Hello, I'm Bill Clinton. What's your name?
Hello, Bill. I'm Nelson Mandela. Where are you from?
I'm from the United States of America.

3 Write the names of the people you meet and where they come from.

2 *This is Bruno and Maria*

Present simple (2): *to be;* **possessive adjectives; articles (1):** *a/an*

He's a journalist.

He's an _____.

She's a police officer.

She's a _____.

She's an artist.

He's a _____.

VOCABULARY AND SOUNDS

1 Complete the sentences with words from the box.

artist police officer receptionist secretary teacher
waiter doctor engineer student farmer journalist

2 Turn to Communication activity 10 on page 100 and check you know what the other jobs are.

3 🔊 Listen and look at the stressed syllables.

<u>ar</u>tist <u>jour</u>nalist <u>sec</u>retary rec<u>ep</u>tionist
engi<u>neer</u>

4 🔊 Listen and repeat these words.

teacher waiter doctor farmer student

5 Work in pairs. Underline the stressed syllables in the words in 4.

🔊 Listen and check.

READING

1 Look at the photos. Where do you think the people are from, *Britain, Brazil* or *Japan*?

2 Match the photos and the descriptions below them.

3 Answer the questions.

1 Where is Pete from?
2 What's his job?
3 Where are Bruno and Maria from?
4 What's their job?
5 Where is Michiko from?
6 What's her job?

A This is my friend Pete Jenkins. He's from London in Britain. He's an engineer.

B This is Bruno and Maria Rodrigues. They're from São Paulo in Brazil. They are doctors.

C This is my friend Michiko. She's from Tokyo in Japan. She's a teacher.

GRAMMAR

Present simple (2): *to be*

I'm (= I am)
you're (= you are)
he's (= he is) *she's* (= she is) *it's* (= it is)
we're (= we are)
they're (= they are)

Possessive adjectives

	my	
What's	your	
	his	name?
This is	her	friend.
	our	
	their	

Articles (1): *a/an*

You use *a/an* before jobs.
*I'm **a** student. He's **an** engineer.*

You use *a* before all nouns which begin with a consonant.
*What's your job?   I'm **a** student.*

You use *an* before all nouns which begin with a vowel.
*What's his job?   He's **an** accountant.*

1 How many different forms are there for the present simple of the verb *to be*?

2 Match the questions and the answers.

1	Where's he from?	a	Her name's Fatima.
2	What's her name?	b	She's a doctor.
3	Where are they from?	c	His name's Andrew.
4	What's his name?	d	He's from Brazil.
5	What's her job?	e	They're from Argentina.

3 Choose the correct sentence.

1 a His from Brazil.
   b He's from Brazil.
2 a Their from Argentina.
   b They're from Argentina.
3 a He's name is Andrew.
   b His name's Andrew.
4 a Their name's Rodrigues.
   b They're name is Rodrigues.

4 Complete the sentences with *a* or *an*.

1 He's _____ engineer.
2 I'm _____ student.
3 She's _____ artist.
4 She's _____ doctor
5 He's _____ teacher.
6 He's _____ waiter.

WRITING

1 Write a description of the people in the chart. Say who they are, where they're from and what their job is.

Name:	Petra Schmidt	Rosario Barbisan
Town/country:	Hamburg Germany	Montevideo Uruguay
Job:	Student	Teacher

This is Petra Schmidt.
She's from ...

2 Think of someone you know. Write a description. Say who they are, where they're from and what their job is.

(3) Questions, questions

Questions; negatives; short answers

LISTENING AND SPEAKING

1  Read and listen to the first part of this conversation.

JANIE   Hi, Holly, how are you?
HOLLY   Fine thanks, how are you?
JANIE   I'm OK. Who's this?
HOLLY   This is Greg.
JANIE   And how old is he?
HOLLY   He's twenty-three.

2 Put the sentences a – e in the second part of the conversation.

JANIE   What's his job?
HOLLY   (1) _a_
JANIE   An actor! What's his surname?
HOLLY   (2) _c_
JANIE   Is he from Hollywood?
HOLLY   (3) _d_
JANIE   He's very good-looking. What's his phone number?
HOLLY   (4) _b_
JANIE   Is he your boyfriend?
HOLLY   (5) _e_

a He's an actor.   b That's a secret!   c Sheppard.
d No, he isn't. He's from New York.   e Yes, he is.

Listen and check.

3 Work in pairs. Act out the conversation.

GRAMMAR

> ### Questions
> **You can form a question:**
> – with a question word.
> **What**'s his job?   **How** old is he?
> – without a question word.
> **Is** he from New York?   **Are** you an actor?
>
> ### Negatives
> **I'm not** (= I am not)
> you are**n't** (= you are not)
> he is**n't** (= he is not) she is**n't** (= she is not)
> it is**n't** (= it is not)
> we are**n't** (= we are not)
> they are**n't** (they are not)
>
> ### Short answers
> Is he your boyfriend?   Yes, he is.
> Is he from Hollywood?   No, he isn't.
> Are you an actor?   No, I'm not.
> Is your surname Sheppard?   No, it isn't.

1 Put the words in the right order and write sentences.

1 old you are how?
2 you a teacher are?
3 her what's surname?
4 what's job your?
5 are children how your?
6 her what's address?

2 Look at the conversation in *Listening and speaking* activities 1 and 2. Are these sentences true or false? Write T for true and F for false.

1 Greg isn't a doctor.
2 He isn't from New York.
3 He isn't twenty-four.
4 He's from Hollywood.
5 His surname is Sheppard.
6 His phone number is a secret.

3 Match the questions and the answers.

1 Is he American?   a No, it isn't.
2 Is his first name Pete?   b Yes, he is.
3 Are you a student?   c Yes, they are.
4 Are they friends?   d Yes, I am.
5 Is she from New York?   e No, she isn't.

4 Write short answers to the questions.

1 Are you from England?   4 Are you sixteen?
2 Are you married?   5 Are you a student?
3 Is your surname Smith?   6 Is your phone number 01278 66554?

1 No, I'm not.

VOCABULARY AND WRITING

1 Complete the missing information on the Landing Card with words from the box.

> married   surname   address   age   job   first name   phone number

Landing Card

1 _____	SHEPPARD
2 _____	GREG
3 _____	ACTOR
4 _____	23
5 _____	365 AVENUE OF THE AMERICAS, NEW YORK N.Y 10021
6 _____	(212) 693-4428
7 _____	NO

2 Look at the Landing Card in 1 and complete this description.

____ Sheppard is an ____ and he's ____. His address is 365 Avenue of the Americas, New York, and his ____ is (212) 693-4428. He isn't ____.

3 Look at the Landing Card below.

Landing Card

1	Surname	NORTH
2	First name	FIONA
3	Job	TEACHER
4	Age	25
5	Address	17, HILL STREET, BRISTOL
6	Phone number	0117 56731
7	Married	YES

Now look at the description in 2 again and notice how you join two pieces of information together with *and*.

*Greg Sheppard is an actor **and** he's twenty-three.*

Now write a description of Fiona North. Use the description in 2 to help you. Join two pieces of information together with *and*.

4 | *How many students are there?*

There is/are; plurals (1): regular; position of adjectives

SOUNDS AND VOCABULARY

1 🔊 Listen and repeat these numbers.

1 2 3 4 5 6 7 8 9 10 11
12 13 14 15 16 17 18 19 20

2 Write the numbers for these words.

> three thirteen thirty four fourteen
> forty five fifteen fifty six sixteen
> sixty seven seventeen seventy
> eight eighteen eighty nine
> nineteen ninety one hundred

Now check your answers.
Turn to Communication activity 20 on
page 104.

3 Look at the stress pattern of these words.

☐	☐ ☐	☐ ☐
three	thirteen	thirty
four	fourteen	forty

Match these words with the stress patterns above.

five   fifteen   fifty
six   sixteen   sixty
seven   seventeen   seventy
eight   eighteen   eighty
nine   nineteen   ninety

🔊 Now listen and check.

4 Look at the words in the box. Which of the items can you see in the photo?

> classroom   self-study room   café
> reception desk   computer
> cassette player   book   table   chair
> car park   library

5 Look at the adjectives in the box.

> friendly beautiful international
> good kind interesting
> popular

Which adjectives can you use to describe:
a person, a school, a classroom, a town, a lesson?

a person – friendly, beautiful ...

READING

1 Read the brochure for the International English School. Find the answers to these questions.

1 Where's the International English School?
2 Why is it 'international'?
3 How many teachers are there?
4 Where are the students in Kevin's class from?
5 Who is Patricia's teacher?
6 What's the phone number?

2 Work in pairs and check your answers.

Welcome to the International English School in Oxford!

At our school there are two hundred students from thirty countries. There are twelve students in a class, twenty classrooms and thirty teachers. In each modern classroom there are tables, chairs and a cassette player. There are two self-study rooms with computers and a library. There's a café and a car park.

Our students say:

'In my class there are two people from Italy, three from Brazil, five from Japan and one from Greece. It's a very international class!' *Kevin, Malaysia*

'Andrew is a good teacher and he's very popular.' *Patricia, France*

'Oxford is a lovely city. The colleges are beautiful and the people are very kind.' *Nourredine, Morocco*

'The lessons are very interesting.' *Efstathia, Greece*

For more information, here's our address and phone number:

5, George Street, Oxford, England 01865 27462

Or come and visit us. The people at the reception desk are very friendly!

GRAMMAR

> *There is/are*
> **There's** a door. (= There is a door.)
> **There are** seven chairs.   **Is there** a car park?
>
> Plurals (1): regular
> **You add -es to form the plural of nouns which end in -s:**
> *class – classes*   *address – addresses*
>
> Position of adjectives
> **You can put an adjective in two positions:**
> **– after the verb** *to be.*
> *She's **kind.**   Oxford **is lovely.***
> **– before the noun.**
> *She's a **kind** woman.   Oxford is a **lovely** city.*

1 Complete the sentences with *there is* or *there are*.

1 You use ____ when the noun is singular.
2 You use ____ when the noun is plural.

2 Singular or plural?

1 There is a *computer/computers* in the two self-study *room/rooms*.
2 There are *table/tables* and *chair/chairs*, and a cassette *player/players*.
3 There are thirty *teacher/teachers*.
4 The *college/colleges* in Oxford are beautiful.
5 There is one *student/students* from Greece.

3 Put these words in the right order and write sentences.

1 very classrooms are comfortable the
2 Oxford a city is beautiful
3 the is international class very
4 lessons are interesting the very
5 is teacher friendly their
6 the Oxford are kind people in

SPEAKING AND WRITING

1 Work in pairs. Write a brochure for your ideal school. Say:

– what the school is called
– where your school is
– the number of students there are
– the number of teachers there are
– what there is in the school
– what there is in each classroom
– what the address and phone number is

Use the brochure in *Reading* activity 1 to help you.

2 Put your brochures on the classroom wall for other students to read.

5 | *Where's my pen?*

***Has/have got*; prepositions of place (1)**

LISTENING AND VOCABULARY

1 Complete the conversations below with these words.

It's chair Where's Thanks is have

A (1) _____ my pen?
B (2) _____ on the table, near your book.
A Oh, I see. Thanks.

A Have you got a mobile phone?
B Yes, I (3) _____. It's in my bag. Here you are.
A (4) _____.

A Where's my bag?
B What colour (5) _____ it?
A It's blue.
B It's under your (6) _____.
A Oh, yes. Thank you.

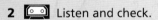

2  Listen and check.

3 Go round and act out the conversations with other students in the class.

4 Match the words in the box with the items in the photos.

> mobile phone calculator pen glasses watch camera
> personal stereo keys comb wallet diary ring bag
> pencil book notebook coat

5 Match the words in the box and the colours.

> red orange yellow pink blue purple green
> brown white grey black

6 What colour are the items in the photo?

pink bag ...

SOUNDS

1  Listen and repeat the alphabet.

a b c d e f g h i j k l m
n o p q r s t u v w x y z

2 Put the letters of the alphabet in the correct column.

/eɪ/ /iː/ /e/ /aɪ/ /əʊ/ /uː/ /ɑː/
a b f i o q r

 Now listen and check.

GRAMMAR

обладание
владение

> **Has/have got**
> **You use *have got* when you talk about possession.**
> ***Have got* means the same as *have*. You use it in spoken and informal written English.**
> *He's got/has got a pen. (= He has a pen.)*
> *They've got/have got a new car. (= They have a new car.)*
> ***Have you got** a watch? Yes, I have. No, I haven't.*
>
> Prepositions of place (1)
> *He's got a mobile phone **in** his bag. The pen is **on** the table.*
> *The bag is **under** the chair. His pen is **near** his book.*

1 Work in pairs. Say two or three things you've got in your bag.

I've got a ...

2 Work with another student.

Student A: In turn, ask Student B what he/she's got in his/her bag. Score 1 point for every *Yes* answer. Answer Student B's questions.

Student B: Answer Student A's questions. In turn, ask Student A what he/she's got in his/her bag.

Have you got a calculator?   Yes, I have.
Have you got a pencil?   No, I haven't.

3 Look at the picture. Which of the following statements are true?

1 The bag is on the table.
2 The mobile phone is under the chair.
3 The book is under the table.
4 The pen is in the bag.
5 The calculator is near the glasses.
6 The coat is on the chair.

4 Go round asking and answering:

– what someone has got in their bag
– where something is

What have you got in your bag?
Guido, where's your bag?

SPEAKING AND LISTENING

1 Work in pairs. Make a list of ten possessions a typical teenager has got in your country.

2 📼 Listen to Steve from Britain. Tick (✓) the things that he's got.

radio television bicycle watch personal stereo video
mobile phone   computer

Did he mention any of the things you wrote in 1?

3 Work in pairs. Try to remember as much as possible.

📼 Now listen again and check.

4 Turn to Communication activity 2 on page 99.

5 Work in pairs. Try to remember what there is in the picture.

Progress check 1–5

VOCABULARY

1 Write new words in a notebook. Write down their meaning in your language and the lesson they come from.

married – marié (Lesson 3), book – livre (Lesson 4)

You can also think of topic headings to put them under.

Jobs – *journalist   police officer   secretary*
Possessions – *pen   watch   keys   diary*

Here are some words from Lessons 1 to 5.

American   age   book   blue   diary   doctor
Hungary   Korean   female   red   pen
personal stereo   taxi driver   Thailand   first name

Write the words down under one of the topic headings below. (Some words can go under more than one heading.)

countries   nationalities   jobs   classroom   colours
personal possessions   personal information

Now look back at Lessons 1 to 5 and write down words you want to remember.

2 Here are some questions and replies to use when you have difficulty understanding something.

What's *cartolina* in English?   I don't know.
What does *popular* mean?   I'm not sure.
How do you say *nom de famille* in English?
                                                        I don't understand.

Choose five or six new words from Lessons 1 to 5 which you don't understand, and five or six words in your language which you'd like to know in English.

Now go round the class asking what they mean or what they are in English.

What does 'favourite' mean?   I'm not sure.
How do you say 'ami' in English?   Friend.

GRAMMAR

1 Put the words in the right order and write sentences.

1 name's Peter my.   4 there students how are many?
2 you married are?   5 your name what's?
3 are you how?   6 pen on the table is.

1 My name's Peter.

2 Write the possessive adjectives which go with these subject pronouns.

I   you   he   she   we   they

I – my

3 Choose the correct word.

1 *She/her*'s from England.   4 *He's/his* name is Phil.
2 *We/our*'re students.   5 *Your/you*'re a student.
3 *I/my* age is twenty-two.   6 *They/their* coats are on
                                                        the table.

4 Here are some answers. Write the questions.

1 My name's Andrew.   4 Her name's Lah.
2 Britain.   5 She's from Bangkok.
3 I'm an artist.   6 She's Thai.

1 What's your name?

5 Look at the photo. Write the answers to the questions.

1 What's his name?
2 How old is he?
3 What's his job?
4 Where's he from?

1 His name is Paul Harris.

Name   Paul Harris
Date of birth   20/4/76
Address   London
Occupation   Student

6 **Answer these questions with short answers.**

1 Is your teacher in the classroom?
2 Are there twenty tables?
3 Are there ten students?
4 Is there a cassette player?
5 Is it a small classroom?
6 Are you a student?

1 Yes, she is.

7 **Look at the picture. Write sentences saying where the things are.**

The bag is on the table.

SOUNDS

1  **Listen and tick (✓) the words you hear.**

1 hit   heat
2 sit   seat
3 tick   teak
4 fit   feet
5 rich   reach
6 pip   peep

Now say the words aloud.

2  **Listen to the stress pattern in these words.**

□ □   England   Poland   Thailand
□ □   Brazil   Japan   Peru
□ □ □   Germany   Mexico   Hungary

Now say the words aloud.

3 **Match these words with the stress patterns in 2.**

mother   teacher   police   cassette
noisy   stereo   camera   trainers

 **Now listen and check.**

READING AND WRITING

1 **Match the questions 1 – 3 and the answers a – c about English. Use a dictionary, if necessary.**

1 How many people speak English?
2 How many people speak English as a second language?
3 How many words are there in English?

a Another 350 million people, in countries such as Nigeria, Kenya, India and Pakistan.
b There are about 1 million words in English.
c 350 million people speak English as a first language, in countries such as Britain, United States of America, Canada and Australia.

2 **Look at the punctuation in the questions and answers in 1. When do you use:**

a a capital letter   b a comma   c a question mark   d a full stop

Can you think of other occasions when you use a capital letter in English?

3 **Rewrite these three questions and three answers about English with capital letters, commas, question marks and full stops.** *får, fören*

how many people speak english as a foreign language 100 million people speak it as a foreign language in countries like france italy brazil and thailand is there any other language with more speakers there are about 1 billion speakers of chinese as a first language how many words do people use in everyday speech most people only use about 10000 words

6 | *Families*

Possessive *'s* and *s'*; plurals (2): regular and irregular

Holly

VOCABULARY AND SOUNDS

1 Look at the photo of Holly and her family. Can you guess who the people are? Use the words in the box to help you.

> brother father mother sister uncle grandfather
> grandmother aunt husband wife son daughter
> girlfriend boyfriend

2 Turn to Communication activity 14 on page 102 to check your answers.

3 Work in pairs. Say who the people are in relation to Holly.

Jenny is her mother.

4 🔲 Listen to these words. The underlined sound is /ə/.

brother father mother sister

5 Underline the other words in the box in 1 with a /ə/ sound.

🔲 Listen and check. As you listen, say the words aloud.

GRAMMAR

pazésiv притяжательный падеж

> ### Possessive *'s* and *s'*
> **You add the possessive *'s* to singular nouns for people instead of *of* to show possession.** pa'ze SHən
> *My husband's name is Philip.*
> not *The name of my husband is* ...
> **You add ' to plural nouns ending in *-s*.**
> *My sons' names are Andrew and Steve.*
> not *The names of my sons are* ...
> **Remember:**
> *My father's name is Jack. (= The name of my father is Jack.)*
> *My father's a doctor. (= My father is a doctor.)*
>
> ### Plurals (2): regular
> **For nouns which end in *-y* you drop the *-y* and add *-ies*.**
> *family – families*   *baby – babies*
> **For other regular plural endings, see Lesson 4 on page 9.**
> ### Irregular
> **There are many irregular plurals in English. Here are some of them.**
> *child – children*   *man – men*   *woman – women*
> *person – people*

1 Look at these sentences. Are they possessive *'s* or contraction *'s*? сокращение

1 Holly's mother is Jenny.
2 Andrew's married.
3 Jenny's husband is Philip.
4 Her children's names are Andrew, Steve and Holly.
5 Holly's family is quite big.
6 Antonia's her brother's wife.

2 Put the ' in the right place.

My parents names are Michel and Paulette. My mothers a journalist and my fathers unemployed. Ive got two brothers and a sister. My brothers names are Pierre and Thierry. My sisters names Patricia. Shes married to Jean. Shes a teacher and hes a banker.

3 Explain the meaning of these words.

aunt   uncle   cousin   grandfather   grandmother

aunt – my mother's sister or my father's sister

4 Work in groups of four or five. Write a list of your relatives' names. In turn, ask and say who each person is.

READING AND SPEAKING

1 You're going to read about Kibiri and his family. Look at the photo. Where do you think Kibiri lives?

2 Check the meaning of the underlined words in your dictionary.

They <u>live</u> with Kibiri's father.
They all <u>work</u> on the farm.
The women and the girls <u>stay</u> at home with the old and the sick people.
The women <u>make</u> yoghurt and <u>sell</u> it for food.
Kibiri's wife and their daughters <u>walk</u> ten kilometres ...

3 Read the passage and look at the photo. Who is Kibiri?

4 Work in pairs. Look at the photo and say who you think the people are.

5 Work in pairs. Describe a typical family in your country.

In a typical family in my country there are two or three children.

Kibiri is a farmer and a vet in Senegal. He's thirty-five years old. Kibiri and his wife have got seven children, five girls and two boys. They live with Kibiri's father and his brother's family. They all work on the farm. The men and the older boys are often away from home with the animals. The women and girls stay at home with the old and sick people. The women make yoghurt and sell it for food, oil, salt and other things for the home. Kibiri's wife and their daughters walk ten kilometres twice a day to get water. It's hard work but they're a very happy family.

What's the time?

Present simple (3) for customs and routines; prepositions of time (1)

It's one o'clock.

SPEAKING AND VOCABULARY

1 Look at the clocks and listen to the times.

It's five past twelve.   It's twenty to seven.   It's a quarter past one.   It's half past seven.   It's a quarter to five.

Now write the times for the clocks below.

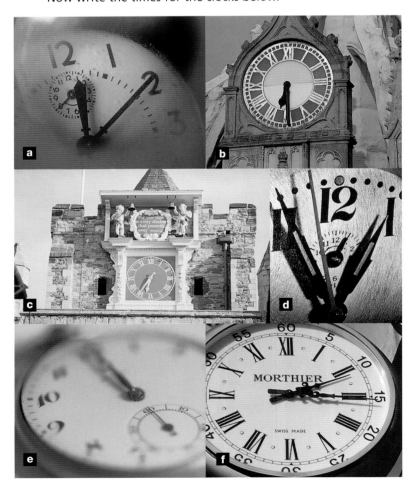

a
b
c
d
e
f

2 Work in pairs and check your answers to 1.

3 Write down three different times.

Now work in pairs. Ask and say what time it is.

Excuse me, what's the time?
It's five to seven.

4 Look at the words in the box. Find:

– five times of the day
– three meals
– two things you do every day

afternoon breakfast dinner evening finish get up
go to bed go to work/school lunch midday
morning night start weekend work

Complete the sentences with words from the box.

1 You have _____ in the morning.
2 You have dinner in the _____.
3 You _____ at night.
4 People _____ work at five in the afternoon.
5 Many people don't work at the _____.
6 You have _____ at midday or in the afternoon.
7 You get up in the _____.
8 People start _____ at half past eight in the
   _____.

READING AND LISTENING

речи

1 Read *Daily routines around the world* and decide which statements are true for your country.

всё не распорядок

Daily routines around the world

- ☐ In Austria children go to school at half past seven in the morning.
- ☐ In Germany people go to work between seven and nine in the morning.
- ☐ In Holland people start work at eight in the morning and finish work at five in the afternoon.
- ☐ In Greece children start school at eight and finish at one-thirty or start at two and finish at seven in the evening.
- ☐ In France people have lunch at midday.
- ☐ In Spain people have lunch at three or four o'clock in the afternoon.
- ☐ In the USA people finish work at five in the afternoon.
- ☐ In Norway people have dinner at five in the afternoon.
- ☐ In Spain people have dinner at ten or eleven in the evening.

сравни

2 Work in pairs and compare your answers.

3 📼 Listen to Tony, from Australia. Tick (✓) the statements in *Daily routines around the world* which are true for Australia.

ō

GRAMMAR

общае в распорядок.

Present simple (3) for customs and routines

You use the present simple to talk about customs and routines.

In Spain people **have** *dinner at ten or eleven in the evening.*

The form of the present simple is the same for all persons except the third person singular (*he/she/it*). (For more information on the third person singular form, see Lesson 9.)

I	
You	*leave work at five in the afternoon.*
We	*start work at nine.*
They	

You form the negative with *don't*.

The Australians **don't** *have lunch at midday.*
They **don't** *have dinner at five.*

Prepositions of time (1)

in: *in the evening   in the morning*
at: *at night   at midday   at midnight   at seven o'clock*
      at the weekend

1 Rewrite the statements in *Daily routines around the world* so they are true for your country.

We don't start work at eight. We start at nine.

2 Complete these statements about Australia with *a/an*, *the* or put a – if no article is needed.

1 In Australia we get up at seven in _____ morning.
2 We have lunch at _____ one o'clock.
3 We stop work at five in _____ afternoon.
4 We go to bed at eleven or twelve at _____ night.
5 We don't work at _____ weekend.

WRITING

1 Read this informal letter to a friend and answer the questions:

   – how do you start an informal letter?
   – where do you write your address?
   – where do you write the date?
   – how do you finish a letter?

```
                      13, Hill Top Road
                      Edinburgh
                      12/3/97
Dear Francesca,

Thank you for your letter about the
times you do things in Italy. In
Britain, we get up at seven or eight
o'clock in the morning. We have
breakfast at eight o'clock and then we
go to work. We work from nine in the
morning to five in the afternoon and
then we go home. We have dinner at
six or seven o'clock in the evening,
and we usually go to bed at eleven
o'clock or midnight.

Best wishes,
                  James
```

2 Write a letter to a friend. Say what time you do things in your country.

8 | *Home*

Some and **any** (1); prepositions of place (2)

The Kapralov Family, Sudzal

Number of people in the house: 4
Living area: 140m2
Household equipment: 2 radios, 1 stereo
system, 2 telephones, 2 televisions, 1 car
Most treasured possession: video games,
Barbie doll
Percentage of income spent on food: 25%

VOCABULARY AND SOUNDS

1 Tick (✓) the rooms and places there are in your home.

> sitting room dining room kitchen garden bathroom
> bedroom study garage

Now work in pairs. Ask and say what rooms and places there are in your home.

Is there a sitting room?
Yes, there is.

2 Work in pairs. In which rooms do you expect to see the following things?

> armchair bath bed bookcase chair cooker
> cupboard curtains dishwasher fire fridge lamp
> sofa table television toilet shower washbasin

3 Look at these words and underline the stressed syllable.

curtains cooker sofa kitchen study
television shower

4 Work in pairs and check your answers to 3.
[cassette icon] Now listen and check.

5 Work in pairs. Look at the box and say where the rooms and places in 1 are in your homes.

> upstairs downstairs in the garden outdoors indoors
> at the back   at the front

Our sitting room is at the back of the flat.

Our dining room is downstairs at the front of the house.

READING

Read the information about the Kapralov family. Are these statements true or false?   *шторн*

1 There are some curtains. *кортнс*
2 There aren't any carpets.
3 They've got some video games.
4 They haven't got any bookcases.
5 There are some cupboards in the kitchen.
6 They haven't got any mirrors.

GRAMMAR   *положит.*
a'farmotiv утвердительни

> ### *Some* and *any* (1)
> **You usually use *some* with plural nouns in affirmative sentences when you're not interested in the exact number.**
> *There are **some** chairs in the sitting room.*
> **You usually use *any* with plural nouns in negative sentences and questions.**
> *There aren't **any** curtains in the bathroom.*
> *Are there **any** cupboards in the kitchen?*
>
> ### Prepositions of place (2)
> *The car is **in front of** the bookcase.*
> *The bookcase is **behind** the car.*
> *The radio is **next to** the television.*
>
> **(For more information on prepositions of place, see Lesson 5.)**

1 Complete the sentences with *some* or *any*.

1 There aren't _____ bookcases in the dining room.
2 They have _____ curtains in the sitting room.
3 There are _____ flowers on the table.
4 They haven't got _____ lamps in the kitchen.
5 Are there _____ carpets in the house?
6 They've got a table and _____ chairs in the study.

2 Choose the correct word.

1 *There/they're* from London.
2 *Their/there* children are at school.
3 *They're/their* not at home.
4 *There/they're* aren't any mirrors.
5 *Their/they're* house is quite large.
6 *There/their* are two doors.

3 Look at the photo of the Kapralov family and say where these things are.

1 bookcase/television
2 plant/lamp
3 window/door
4 chair/table
5 armchair/fire
6 mirror/door

There's a bookcase behind the television.

LISTENING AND WRITING

1 You're going to hear an interview with Geoff, from Scotland, who lives on a boat. Look at the photo. Can you guess the answers to the questions in the chart?

	Geoff	**Your partner**
Type of home	*boat*	
Size		
Number of rooms		
Number of people		
Furniture		
Most important items		

🔲 Listen and check.

2 Work in pairs and complete the chart.
🔲 Now listen again and check.

3 Find out about your partner's home. Complete the chart.

9 |*How do you relax?*

Present simple (4) for habits and routines:
Wh- **questions; third person singular** *(he/she/it)*

VOCABULARY AND READING

1 Match the phrases in the box with the pictures below.

> drink coffee   eat an apple   go running   listen to the radio
> play football   read a newspaper   watch a tennis match
> have a shower   learn the guitar

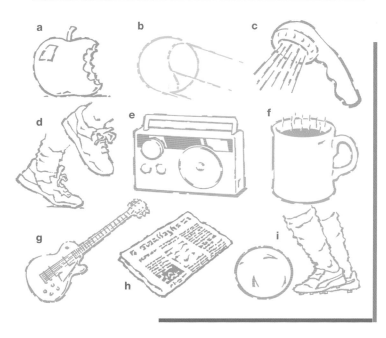

2 Read *How do you relax?* Say who you can see in the photos.

3 Answer the questions.

1 Where does Fiona live?
2 What does Newton do?
3 When does Ingrid go running?
4 Why does Otto get a video?
5 How does Patricia relax?
6 Who does Tanya live with?

4 Write down the verbs in the vocabulary box in 1.

drink, eat ...

Now find phrases in the passage which go with the verbs.

drink tea, eat toast, chocolate ...

How do you relax?

'I go running in the evening, and then I come home and have a shower.' **Ingrid**, Tromso

'My husband goes to his club to relax, and I sometimes go with him.' **Tanya**, Marseille

'I have a hot bath, and listen to the radio.' **Patricia**, Brussels

'I play the guitar, mostly *Rolling Stones* songs.'
Karl Heinz, Vienna

'My girlfriend likes television, so I get a video and we watch it together.' **Otto**, Budapest

'I drink tea and eat toast and jam in the garden.'
Carrie, Manchester

'I eat chocolates while my boyfriend reads novels to me.'
Fiona, Hong Kong

'I learn languages to relax. I'm learning Russian at the moment. My wife learns languages too.'
Gerard, Lourenco Marques

'I never relax. I'm a taxi driver.' **Newton**, Rio

GRAMMAR

инфинитив неопределенная форма (handwritten)

> **Present simple (4) for habits and routines**
> You use the present simple to talk about habits and routines. We usually say *how often* or *when* we do something when describing a routine.
> *I **go** running in the evening and then I **come** home and have a shower.*
>
> *Wh-* questions
> You form *Wh-* questions for all verbs except *to be* with *do* or *does* + infinitive. (For *Wh-* questions with *to be*, see Lesson 3.)
> *What **do** you **do** to relax? Where **does** Fiona **live**?*
>
> Third person singular *(he/she/it)*
> You form the third person singular by adding *-s* to most verbs.
> *drink – drinks   eat – eats   learn – learns*
> *like – likes*
> You add *-es* to *do*, *go* and verbs which end in *-ch, -ss, -sh* and *-x*.
> *do – does   go – goes   wash – washes*
> **(For more information, see Grammar review page 107.)**

1 Find four examples of the third person singular in the passage. *pasij отрывке происхождения* (handwritten)

2 Work in pairs. Read *How do you relax?* again. Ask and say what the people do to relax.

What does Ingrid do to relax?
She goes running.

3 Complete the questions about the passage.
1 What _____ Ingrid _____ in the evening?
2 Where does Tanya's husband _____?
3 _____ music does Karl Heinz _____?
4 Who _____ Otto _____ a video with?
5 _____ does Carrie drink tea and _____ toast?
6 Why _____ Newton never _____?

Now find the answers.

4 Work in pairs. Ask and answer the questions in 3.

5 Work in pairs. Ask and say what you do to relax.

SOUNDS

1 🔊 Listen and repeat these verbs.

/s/   /z/   /ɪz/
gets   does   watches

2 Look at these verbs and put them in the correct columns above.

lives makes finishes washes eats drinks listens
plays reads likes
🔊 Now listen and check.

3 🔊 Listen and tick (✓) the sentences you hear. Do you hear *he* or *she*?
1 a What does he do?   b What does she do?
2 a Where does she live?   b Where does he live?
3 a How does he see it?   b How does she see it?
4 a Where does she go?   b Where does he go?

LISTENING AND SPEAKING

1 Tick (✓) the things you do in your free time or when you relax.

	You	**Helen**	**Chris**
go to the cinema			
watch television			
listen to the radio			
do some sport			
learn a language			
play music			
read a novel			
see friends			
go to a club			

2 🔊 Listen to Helen and Chris. Tick (✓) the things they do in their free time.

3 Work in pairs and check your answers to 2.

Helen sees friends and goes to the cinema.

4 🔊 Listen again and check.

Do you like jazz?

Pronouns; present simple (5): talking about
likes and dislikes

VOCABULARY AND READING

1 Tick (✓) the things in the box that you like.

football   cooking   tennis   magazines   going to parties   swimming
classical music   dancing   jazz   going to the theatre   going to the cinema
watching sport   rock music   Chinese food

2 Put the words and phrases in 1 in two columns: *nouns* and *words or phrases
ending in -ing.*

nouns: football   *words or phrases ending in -ing: cooking*

3 Read the adverts and find an event for someone who likes:

jazz music   Chinese food   ballet   films   magazines

4 Write down information about the events you chose in 3.

Name				
Type				
Date/days				
Time				
Place				
Other details				

LONDON FESTIVAL BALLET
performing
Swan Lake
at the Coliseum
Friday 26th, Saturday 27th August
Tickets £15.00, £25.00

Gil Scot-Heron
12th/13th/14th September
Jazz Cafe
Camden Town

Jam session of Irish a... music is at the Ship Hotel, Place.

Warner Brothers
Multiplex
CINEMAS
now showing:

Twister
Kids
Richard III
The Hunchback of Notre Dame

LISTENING

1  Listen to the conversation. Underline anything which is different from what you hear.

A Do you like classical music?
B No, I don't. I hate it.
A What kind of music do you like? Do you like rock music?
B Yes, I do. I love it.
A Who's your favourite singer?
B Mick Hucknell from *Simply Red*. He's great.
A I don't like him very much. I don't like *Simply Red*.
B Oh, I like them very much. What about you? What's your favourite music?
A I like jazz. My favourite musician is Stan Getz. Do you like jazz?
B It's all right.

2 Work in pairs and correct the conversation in 1.
 Now listen and check.

3 Work in pairs and act out the conversation.

GRAMMAR AND FUNCTIONS

месторешение

> **Pronouns**
> *I* like Simply Red. I like **them** very much.
> *He likes Stan Getz but I don't like **him** at all.*
>
> **Subject: *I   you   he   she   it   we   they***
> **Object: *me   you   him   her   it   us   them***
>
> **Present simple (5): talking about likes and dislikes**
> *What kind of music do you like?   I like classical.*
>   *I don't like rock.*
> *Do you like jazz?   Yes, I do.   I love it.*
>   *Yes, it's great.   It's all right.*
>   *No, I don't like it very much.*
>   *No, I hate it.*

1 Look at the pronouns in the grammar box. How many object pronouns have the same form as their subject pronouns?

2 Complete the sentences.

  A Do you _____ the actress Demi Moore?

  B Yes, I do. _____ great. What about you? Do you like her?

  A No, I _____. I don't like _____ at all.

  B Who's your _____ film star?

  A _____ like Jodie Foster.

3 Choose the correct pronoun.

  1 John likes Sharon Stone but Patti doesn't like *she/her* at all.

  2 My sister hates Stone Roses but I like *they/them* a lot.

  3 *We/us* both like jazz.

  4 My favourite singer is George Michael, but my brother doesn't like *he/him* at all.

4 Write down the names of your favourite:

  film star   singer   sportsperson   politician
  club   TV programme   film   restaurant

5 Find out if other students like the people or things you wrote in 4.

SOUNDS

1 🔊 Listen to the conversation in *Grammar and functions* activity 2. Which speaker sounds interested?

2 Work in pairs and act out the conversation. Try to sound interested.

LISTENING AND WRITING

1 Look at the adverts for penfriends. Is there anyone who likes the same things as you?

> **My name's Octavio** and I'm from Recife in Brazil. I like playing football on the beach, dancing and playing computer games. Write to me.
>
> **I'm Tomasz** and I'm from Wroclaw in Poland. I like ice skating, going to the cinema and photography. I also like watching sport. What about you?
>
> **I'm Nancy** and I live in Seattle. I'm a high school student and my favourite group is *REM*. I really like the singer. I also like playing volleyball, going to the theatre and skiing. Tell me about yourself.

2 🔊 Listen and choose a penfriend for Octavio, Tomasz and Nancy.

3 Write an advert for yourself. Use the adverts in 1 to help you. Say:

  – who you are   – where you're from   – what you like doing

Progress check 6–10

VOCABULARY

1 When you write a new word in your notebook, check you know its part of speech. Here are some of them.

noun (n) – *brother father girlfriend film*
verb (v) – *have finish start*
adjective (adj) – *favourite popular comfortable*
pronoun (pr) – *he she it*
preposition (prep) – *in under*

What parts of speech are the following words?

breakfast   from   husband   on   read   football
restaurant   to   sitting room   start   telephone
red   beautiful

2 It's also useful to write down words and expressions which go with verbs.

have	a shower, breakfast
go	to the cinema, to work
play	tennis, football
listen to	music, the radio
watch	football, the television
leave	work, school

Match the verbs on the left with these words and expressions. There may be more than one possibility.

to bed   dinner   a video   a concert   the guitar   tennis

3 A good dictionary helps you learn more about a new word. Look at this dictionary extract and find out which of the following it shows:

meaning   pronunciation   word stress   part of speech
how the word is used

breakfast /ˈbrekfəst/ *n* breakfast is the first meal of the day which most people eat early in the morning.

EG I get up at eight o'clock and have breakfast.

Look at your own dictionary and find out what features it shows.

GRAMMAR

1 Write the plural form of these words.

1 child   2 boy   3 man   4 woman
5 brother   6 family

1 children

2 Put the ' in the right place.

Ive got two brothers. My brothers names are Tom and Henry. Toms a doctor and Henrys a teacher. Toms married and his wifes name is Jean. Theyve got two children. Henry isnt married but hes got a girlfriend. Her names Philippa.

I've got two brothers.

3 Put these words in the right order to make sentences.

1 in shops Britain at close five-thirty.
2 get up seven I o'clock at.
3 Greece start at eight children in school.
4 the at USA start people work in nine.
5 Holland in have at they dinner six.
6 we the evening go bed at eleven in to.

1 In Britain shops close at five-thirty.

4 Write sentences saying what time you do these things.

1 get up   4 have lunch
2 have breakfast 5 leave work/school
3 leave home   6 have dinner

1 I get up at seven in the morning.

5 Write the third person singular of the present simple tense of these verbs.

1 do   2 like   3 go   4 make   5 play   6 eat
7 watch   8 learn

1 do – does

24

6 Look at the picture and give short answers to these questions.

1 Is there a television?
2 Are there any curtains?
3 Is there a table?
4 Are there any chairs?
5 Is there a lamp? *кн. шкаф*
6 Are there any bookcases?

1 Yes, there is.

7 Complete the sentences with *some* or *any*.

1 There are _____ curtains in the bedroom.
2 There aren't _____ carpets.
3 There are _____ chairs in the living room.
4 Are there _____ armchairs?
5 They haven't got _____ desks.
6 They've got _____ trees in the garden.

8 Complete these sentences with subject or object pronouns.

1 He likes football but I don't like _____ at all.
2 _____ love tennis and so do our children.
3 Jane's favourite film star is Demi Moore, but I don't like _____ very much.
4 My friend Geoff likes classical music and so do _____.
5 My favourite singers are Pavarotti and Domingo, but Jane doesn't like _____.
6 I like her but she doesn't like _____.

SOUNDS

1 Listen and underline the /ə/ sound in these words.

mother   cousin   dinner   listen
television   computer   newspaper
Now say the words aloud. *всух*

подрекрутше

2 Say these words aloud. Is the underlined sound /s/ or /z/?

drink<u>s</u>   Greek<u>s</u>   journalist<u>s</u>   desk<u>s</u>   computer<u>s</u>   pen<u>s</u>   son<u>s</u>
school<u>s</u>   sport<u>s</u>

Listen and check.

3 Listen to these questions. Put a tick (✓) if the speaker sounds interested.

1 What's your name?
2 What's your job?
3 Where's your mother from?
4 What music do you like?
5 What's your favourite group?
6 What time do you get home from work?

Say the questions aloud. Try to sound interested.

READING AND SPEAKING

1 Read the facts about *The age you do things in Britain*. Which ones are true for your country?

The age you do things in Britain

☐ Children start school at five.
☐ Children leave school when they're sixteen or eighteen.
☐ People learn to drive when they're seventeen.
☐ Students go to university when they're eighteen or nineteen.
☐ People leave university when they're twenty-one or twenty-two.
☐ People get married between the age of twenty and thirty.
☐ Men stop work when they're sixty-five.

2 Work in pairs. Talk about the age you do things in your country.

In my country, children start school at six.

11 | *A day in my life*

Present simple (6): saying how often you do things; prepositions of time (2)

READING

стать6а

1 You're going to read a magazine article about Tanya Philips, who is a presenter on breakfast television in Britain. What time do you think her day starts and finishes?

2 Read the article and find out what time her day starts and finishes.

кратцi peueuue

3 Decide where these sentences go in the article.

  a After the programme we always have breakfast and relax.

  b I often go shopping and have lunch with friends.

  c We usually have dinner quite early, at seven o'clock.

4 Write the questions the journalist asked Tanya.

What time do you get up?

GRAMMAR

FrēkwJNSē *частое повторение частото.*

наперiе

Present simple (6): saying how often you do things
You use the present simple and the following adverbs of frequency to say how often you do things.
100% ——————————— 0%
always usually often sometimes never
You usually put the adverb before the verb.
*I **always** get up at seven o'clock.*
*I **sometimes** go shopping in the evening.*
*I **often** have a drink with friends.*
*I **never** do the washing up.* *стираю*
But you put the adverb after the verb *to be*.
*I'm **usually** asleep by nine o'clock.*
спящад
Prepositions of time (2)
***at:** at seven o'clock   at half past three   at the weekend*
***on:** on Sunday   on Tuesday   on Monday morning*
***from ... to:** from Monday to Friday   from seven to nine o'clock.*

My working day

My working day starts very early. From Monday to Friday I get up at half past three and I have a shower and a cup of coffee. I usually leave the house at ten past four because the car always arrives a few minutes early. I get to the studio at about five o'clock and start work. *Good Morning Britain* starts at seven o'clock and finishes at nine o'clock. Then I leave the studio at a quarter past ten. After that, I get home at three o'clock. A woman helps me with the housework and the ironing. I read the newspaper and do some work. Then my husband gets home at half past five in the afternoon and I cook dinner. We stay at home in the evening. We don't go out because I go to bed very early. We usually watch television and then I go to bed at half past eight. I'm usually asleep by nine o'clock. At weekends I don't get up until ten o'clock. In the evening, we often see some friends or go to the cinema. But I'm always up early again on Monday morning.

1 Look at the verbs in the article again. How many are in the negative? How many have *-s* endings?

2 Put the adverbs in brackets in the correct position.

1 I get up at seven o'clock. (usually)
2 I do the ironing. (never)
3 She has a drink with friends. (often)
4 He goes to bed at eleven o'clock. (sometimes)
5 I'm asleep at midnight. (usually)
6 I'm up at five o'clock in the morning. (never)

3 Complete these sentences with a preposition of time.

1 We work _____ Monday _____ Friday.
2 The film starts _____ six o'clock.
3 He leaves home _____ eight o'clock.
4 I do the housework _____ the weekend.
5 She does the ironing _____ Monday.
6 We usually go out _____ Friday evening.

VOCABULARY AND SPEAKING

1 Look at the phrases in the box. Underline the things you always do every day.

go shopping   see friends   do the washing up
do the housework   cook dinner
get the bus/train to work/school/home
do some work/homework   go to the cinema

2 Work in pairs. Ask and say what you do every day.

Do you always go shopping?   *No, I don't.*
Do you always do the washing up?   *Yes, I do.*

LISTENING AND WRITING

1 You're going to hear Sam, who lives in London, talking about a typical day in his life. Look at the things he does. Which ones do you always, usually or sometimes do? At what time do you do them? Which ones do you never do?

Sam

go to a club ☐
go to a party ☐
do some work ☐
play music ☐
have breakfast ☐
have lunch ☐
go to bed ☐
have dinner in a restaurant ☐
go to a concert ☐
meet friends ☐
play football ☐
telephone a friend ☐
go to work/school ☐

2 🔲 Listen and tick (✓) the things Sam does.

3 Work in pairs. What does Sam always, usually and sometimes do? Can you remember at what time he does these things?

4 Work in pairs. You're going to write a magazine article about a typical day in your life. Say:

– what time you get up
I get up at eight in the morning.

– what you do then
Then I usually have some breakfast.

– what you do after that
After that, I usually read the paper.

Use *then* and *after that* to describe when you do things. Remember to join two phrases with *and*.

⑫ *How do you get to work?*

Articles (2): *a/an, the* and no article; talking about travel

The age of the train

Which do you prefer, the plane or the train?

We talk to Katie Francis. She's twenty-eight years old and she's a marketing consultant. She lives in London but she often works in France.

How does she get there?

'I go by train. I take the train from Waterloo station through the Channel Tunnel.'

How long does the train take?

'It takes three hours and I work during the journey. By plane, it takes an hour but you have to get to and from Charles de Gaulle airport.'

What time does she leave London?

'There are several trains a day. I get the train at eight o'clock and arrive at the Gare du Nord station in Paris at midday French time. I work in an office which is very close, so I go there on foot. I'm at work by half past twelve.'

And when does she leave for home?

'I usually stay at a friend's flat, work next day and get the seven o'clock train home. I get to London at nine o'clock British time, and I'm home by a quarter to ten.'

So what's her opinion of the Channel Tunnel service?

'It's great. It's definitely the age of the train.'

VOCABULARY AND READING

1 Work in pairs. Look at the words in the box. Which means of transport can you see in the photo on the opposite page.

> train plane tram bicycle underground ferry car

Now turn to Communication activity 17 on page 103 and check you know what the other words mean.

2 Match the means of transport in 1 with the following places.

> station airport garage bus stop

3 Which verbs go with the different means of transport?

> drive ride take

4 Which is the odd word out?

1 train bus stop station airport
2 ride bike tram
3 car bicycle tram drive
4 underground ferry train airport

5 Match the adjectives with their opposites. Which adjective has no opposite in the box?

> cheap expensive far fast near slow crowded

cheap – expensive

6 Look at the photo of Katie Francis and the title of the article. What do you think *The age of the train* means?

1 Trains are very slow.
2 Trains are very old.
3 Trains are very popular today.

Now read the article and find out if you guessed correctly.

7 Answer the questions.

1 What does Katie Francis do?
2 Where does she live?
3 How does she get to Paris?
4 How long does it take?
5 What time does she leave London?
6 What time does she arrive in Paris?
7 Where does she stay in Paris?
8 Does she like the Eurostar service?

8 What are the advantages of travelling by train for Katie?

It only takes three hours, ...

GRAMMAR AND FUNCTIONS

Articles (2): *a/an*, *the* and no article

You use the indefinite article *a/an*:
– to talk about something for the first time.
*She works in **an** office in Paris.*
– with jobs.
*She's **a** marketing consultant.*
– with certain expressions of quantity.
*There are several trains **a** day.*
You use the definite article *the*:
– to talk about something again.
***The** office is near the Gare du Nord.*
– when there is only one.
***The** Channel Tunnel.*
You don't use an article:
– with most countries.
She goes to France. She lives in Britain.
– with certain expressions.
by train by plane at work at home
(For more information about articles, see Grammar review, page 111.)

Talking about travel
How do you get to work/school?
How long does it take?
How much is it? How far is it?

1 Look for examples of the indefinite and definite article in *The age of the train*. Which rules in the grammar box do they show?

2 Complete the sentences with *a/an* or *the*, or put a – if no article is needed.

Georges Bastien is _____ ticket inspector. He works on _____ Eurostar service between _____ Britain, _____ France and _____ Belgium. He travels to London and Brussels several times _____ week. He lives in _____ house in Paris. _____ house is about thirty minutes from _____ Gare du Nord. And how does he get to _____ work? _____ bicycle!

3 Match the questions and the answers.

1 How do you get to work/school?
2 How long does it take?
3 How much is it?
4 How far is it?
5 Is it crowded?

a Three pounds. It's quite expensive.
b It's about three kilometres.
c No, it isn't.
d It takes about thirty minutes.
e I go by bus.

SOUNDS

1 🔲 Listen to these words. Notice how you pronounce the word *the*.

/ðə/	/ði:/
the train	the airport
the service	the hour
the station	the office
the bus	the underground

When do you pronounce *the* /ðə/ and when do you pronounce it /ði:/?
Now say the words aloud.

2 Put these words in two columns /ðə/ or /ði:/.

the teacher the answer the book the uncle the aunt the evening the afternoon the video

🔲 Listen and check. As you listen, say the words aloud.

SPEAKING AND WRITING

1 Use the questions in *Grammar and functions* activity 3 to find out how other students in your class get to work or school. Make notes.

Fabbio: by bus, ten minutes ...

2 Choose one or two students and write a paragraph using the notes in 1 describing how they get to work or school.

Fabbio goes to school by bus. The journey takes about ten minutes...

13 *Can you swim?*

Can and *can't*; questions and
short answers

VOCABULARY AND READING

1 Match the verbs in the box with the nouns below.

cook do play ride speak use write

a crossword   a computer   a foreign language
a horse   a meal   a musical instrument   poetry
cook a meal ...

2 Here are some verbs which do not always
need nouns after them. Match them with
the photos.

draw ski swim type

3 Read *Are you an all-rounder?* and answer the
questions with *yes* or *no*.

Are you an all-rounder?
Can you ...
1 run 100 metres in 15 seconds?
2 use a computer?
3 cook a meal for six people?
4 swim 500 metres?
5 speak a foreign language?
6 play a musical instrument?
7 ski?
8 draw?
9 ride a horse?
10 write poetry?
11 drive a car?
12 knit a jumper?
13 type quickly?
14 do crosswords?

4 Count your *Yes* answers. Now turn to Communication
activity 26 on page 105 to find out if you're an
all-rounder.

GRAMMAR

Can
Can is a modal verb.
You use *can*:
**– to talk about something you are able to do on most
occasions.**
*I **can** drive.*

Can't
Can't is the negative of ***can*.**
*I **can't** ride a bike.*

**(For more information about modal verbs, see
Grammar review page 111.)**

Questions and short answers
Can you speak English?   Yes, I can.   No, I can't.
**You use *So can I* and *Nor can I* to describe the same
abilities as someone else.**
I can drive. So can I.   I can't cook. Nor can I.

1 Complete the sentences.

1 _____ you swim? Yes, I _____

2 I _____ drive, but I can't cook.

3 I can't ski. _____ can I.

4 I can draw. _____ can I.

5 Can you do crosswords?
   No, _____

6 I can't drive. _____ can I.

2 Decide where the sentences a – d go in this conversation.

JO   What can you do?

PAT (1) _____.

JO   So can I.

PAT (2) _____

JO   Nor can I. But I can speak a foreign language.

PAT (3) _____

JO   Spanish. Can you speak Spanish?

PAT (4) _____

a What language can you speak?

b No, I can't.

c But I can't cook very well.

d I can run a 100 metres in 15 seconds and I can use a computer.

Now listen and check.

3 Work in pairs and act out the conversation.

4 Work in pairs. Ask and say what things in the questionnaire you can do. Use the conversation in 2 to help you.

SOUNDS

1 Listen to the sentences in *Grammar* activity 1. Do you hear /cən/, /cæn/ or /cɑːnt/?

2 Now say the sentences aloud.

SPEAKING AND LISTENING

1 Look at these statements. Which statements can you see in the pictures?

☐ You can see the Great Wall of China from space.

☐ Cats can't swim.

☐ Chickens can't fly.

☐ Computers can write novels.

☐ Cameras can't lie.

☐ England can win the next World Cup competition.

☐ Thin people can't swim very well.

☐ You can never read a doctor's handwriting.

☐ You can clean coins with Coca Cola.

☐ Cats can see in the dark.

2 Work in pairs. Which statements are true?

3 Listen to Ann and Frank discussing the statements. Tick (✓) the ones they think are true.

4 Work in pairs and check your answers to 3.

Now listen again and check.

5 Work in pairs. Write two true statements and one false statement. Use *can/can't*.

6 Work with another pair. Give your statements to them and look at their statements. Try and guess which statement is false.

Max can ski.
Jenny can't drive.
Paco can run 100 metres in nine seconds.

Prepositions of place (3); asking for and giving directions

VOCABULARY AND LISTENING

1 Complete the sentences below with words from the box.

baker	bank	bookshop	car park
chemist	cinema	florist	greengrocer
library	market	newsagent	
phone box	post office	railway station	
restaurant	supermarket	pub	
swimming pool	station	taxi rank	

1 You can buy stamps in a _____.
2 You can park your car in a _____.
3 You can borrow a book from a _____.
4 You can make a phone call from a _____.
5 You can get a taxi from a _____.
6 You can take the train from a _____.
7 You can go swimming in a _____.
8 You can buy bread at a _____.

1 You can buy stamps in a post office.

2 Work in pairs. Look at the vocabulary box and ask and say where you can:

1 buy some flowers
2 take out some money
3 have a meal
4 buy some aspirin
5 get some vegetables
6 have a drink
7 see a film
8 buy a book

1 Where can you buy some flowers?
At a florist.

3 Look at the map and listen to four conversations. Mark the places on the map.

GRAMMAR AND FUNCTIONS

Prepositions of place (3)
*There's a florist **next to** the bank.*
*There's a bank **between** the florist and the post office.*
*There's a bookshop **on** the corner of Queen Street.*
*There's a chemist **opposite** the restaurant.*

Asking for and giving directions
Excuse me, where's the station? How do I get to Queen Street?
Go along Prince Street. Across East Street. Up George Street. Down Valley Road.
Turn left into George Street. Turn right into Queen Street.
It's on the left.   It's on the right.   It's straight ahead.

1 Work in pairs. Check your answers to *Vocabulary and listening* activity 3.

There's a cinema opposite the bank.

2 Look at the map and say where other places are.

There's a taxi rank opposite the swimming pool.

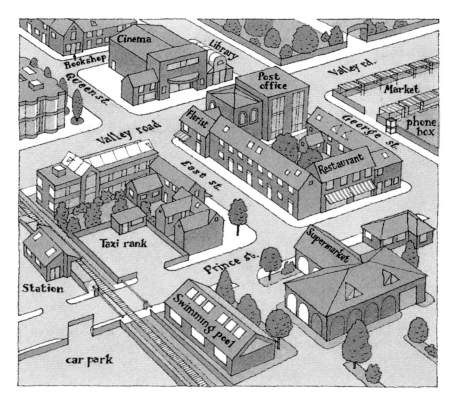

3 Here's a conversation from *Vocabulary and listening* activity 3. Put the sentences in the right order.

a Thank you.
b How do I get to East Street?
c Go along Prince Street. Turn right into East Street. The newsagent is on your left.
d Excuse me, where can I buy a newspaper?
e There's a newsagent in East Street.

[cassette] Now listen and check.

4 Work in pairs. Act out conversations. Use your answers to *Vocabulary and listening* activity 3 and the conversation above to help you.
Excuse me, where can I buy ...?

5 Work in pairs.

Student A: Turn to Communication activity 4 on page 99.

Student B: Turn to Communication activity 12 on page 101.

The end of the High Street?

In the High Street of a British town, there are shops, banks, cinemas and the post office. There's usually the baker's shop, where you can buy bread, the greengrocer's where you can buy fruit and vegetables and the newsagent's, where you can buy newspapers and magazines. In many towns there is also a food market once or twice a week.

In a High Street shop you usually tell an assistant what you want, then you pay the assistant. In a supermarket you choose what you want, and then take it to the check-out and pay there. There is often a queue of people at the check-out.

There are many kinds of restaurants – you often see Italian, Chinese and Indian restaurants. You don't often see people eating outside because it's too cold.

But now there are shopping malls everywhere. They are usually outside the town with all the shops you see in the High Street. You drive there, leave your car in the car park and go shopping. Today shopping malls are very popular. Is this the end of the High Street?

READING AND SPEAKING

1 You're going to read about the High Street, which is often the central street for shopping in a British town. Which of these words do you expect to see?

buy   check-out   assistant   journalist   queue   shopping mall   popular

2 Read the passage. Why is it the end of the High Street? Choose one of the following reasons.

1 Because the shopping malls outside towns are very popular.
2 Because of the food markets once or twice a week.
3 Because there are few assistants in the shops.

3 Read these statements about the passage. Are they true or false for Britain?

1 There are many small shops.
2 There is a market every day.
3 There are assistants to help you.
4 There are often queues in shops.
5 People often eat outside.
6 Many people go shopping in shopping malls.

4 Work in pairs and say if the statements in 3 are true for your country.

15 | *What's happening?*

Present continuous

LISTENING

сообщение
уведомление
объявление
ə'naunsmənt

1 Read the pilot's announcement and decide when it is taking place:
– on take off   – during the flight   – on landing
на взлёте   *во время полёта*   *при посадке*

'**L**adies and gentlemen, this is your pilot speaking. I hope you're enjoying your flight to Milan this morning. At the moment, we're *мы проходим* passing over the beautiful city of Geneva, in the east of Switzerland. If you're sitting on the right-hand side of the plane, you can see the city from the window. We're flying at 10,000 metres and we're travelling at a speed of 700 km/h. I'm afraid the weather in Milan this *дует лёгкий ветерок* morning is not very good. It's raining and there's a light wind blowing. Enjoy the rest of your flight. Thank you for flying with us today. '

2 Listen and underline anything which is different from what you hear.

3 Work in pairs and correct the announcement in 1.

4 Listen again and check your answers to 3.

GRAMMAR

Present continuous *настоящее продолженное время*

You use the present continuous to say what is happening now or around now.

We're flying at 10,000 metres. It's raining in Milan. *настоящее причастие*
You form the present continuous with *am/is/are* + present participle *оформляете*
(verb + -*ing*). (For more information see Grammar review, page 107.)

Questions	Short answers	Negatives
Is she learning to swim?	*Yes, she is. No, she isn't.*	*She isn't learning to swim.*
Are you enjoying the flight?	*Yes, I am. No, I'm not.*	*I'm not enjoying the flight.*

исправленная версию

1 Look at your corrected version of the announcement in *Listening* activity 1 and complete the sentences.

1 The plane _____ going to Rome.
2 They're _____ over Zurich at the moment.
3 They _____ flying at 12,000 metres, they're _____ at 10,000 metres.
4 They aren't travelling at 700 km/h, they _____ travelling at 750 km/h.
5 It _____ raining in Rome. It _____ snowing.

2 Listen to three conversations and decide what the situation is. Choose from the following.

– in an office   – in a restaurant
– in the market   – on a bus
– in the street   – in a gallery
'gæləri

3 Say what the people are doing in the situations in 2. Use the verbs below.

use a computer   have a meal
buy food   go to work
do the shopping   look at paintings *смотреть картины*
In the first conversation, they're having a meal.

34

SOUNDS

1 🔲 Listen and repeat these present participles. *li ing*

having watching lying learning
going shopping buying looking
playing reading talking

2 The stressed words in spoken English are the words the speaker thinks are important. Look at these sentences from the pilot's announcement. Underline the stressed words.

1 This is your captain speaking. *'kaptɜN*
2 I hope you're enjoying your flight.
3 We're passing over the beautiful city of Zurich.
4 You can see the city from the window.

3 🔲 Listen and check. As you listen, say the sentences aloud.

VOCABULARY AND SPEAKING

1 Use the words in the box to label the compass.

> south-west south-east
> north-west north-east

2 Work in pairs. Say where the capital city is in your country.

London is in the south-east of England.

3 Choose words and phrases in the box to describe four of these places.

Marseille New York Moscow Rio Barcelona Bangkok Tokyo
Venice your town

> beautiful attractive on the coast interesting on the river ugly
> in the mountains boring industrial lively lovely modern old
> on an island big small

Marseille – lively, on the coast.

4 Work in pairs. Think of other places you can describe using the words in the box.

READING AND WRITING

1 Read the postcard and answer the questions.

– who's writing the postcard?
– where are they staying?
– where's she writing it?
– who's she writing it to?
– are they enjoying themselves?

Dear Fiona,

Here we are on the island of Mykonos in the south Aegean sea, and we're having a wonderful time. We're staying in a villa in the mountains, near a beautiful beach. I'm writing this postcard in the hotel you can see in the picture. Jacqui is learning to swim, and Terry is lying in the sun at the moment. We're enjoying the holiday very much.

See you soon!

Love, Janet and the kids

Fiona Graham
22, Park Street
Stow-on-the-Wold
Gloucestershire
ENGLAND

2 Work in pairs and check your answers to 1.

3 Write a postcard to a friend. Use the postcard in 1 to help you.
Say:
– where you are
– where you're staying
– where you're writing the card
– what you're doing
– if you're enjoying yourself

Progress check 11–15

VOCABULARY

1 There are many international words in English. Here are some types of international words.

food – *burger pizza*

places of entertainment – *theatre cinema*

types of entertainment – *video music*

sports – *football tennis*

jobs – *president doctor*

brand names – *Coca Cola McDonalds*

Are these words also in your language?

2 Put these international words in the types shown in 1.

Pepsi Toyota secretary restaurant pasta handball

3 There are some words in English which come from other languages.

> siesta concerto spaghetti café judo ballet samba sauna

Are these words also in your language? What languages do you think they come from?

4 Here are some expressions which you use in everyday situations. Match them with the pictures.

> Please   Thank you   Excuse me   Pardon   Sorry

5  Listen and check what people say.

GRAMMAR

1 Complete the sentences with *a/an* or *the*, or put a – if no article is needed.

1 I'm _____ doctor. What do you do?

2 She's French but she lives in _____ Britain.

3 I go to work by _____ train.

4 _____ office where I work is near here.

5 _____ Alps are in western Europe.

6 I work at _____ home.

2 *A or an?*

teacher writer American Hungarian chair ID card aunt uncle novel airport

3 Put the adverb in brackets in the correct position.

1 I get up at nine on Saturday and Sunday. (usually)

2 I see friends on Saturday evening. (often)

3 I do the shopping on Saturday morning. (always)

4 I go for a walk on Sunday. (sometimes)

5 I watch the football in the park on Sunday morning. (often)

6 I go to bed early on Sunday night. (usually)

1 I usually get up at nine on Saturday and Sunday.

a   b   c   d

4 The sentences in 3 are answers to questions. Write the questions.

What time do you get up on Saturday and Sunday?

5 Write sentences saying what Peter can/can't do.

1 play the piano (no)
2 drive (yes)
3 use a computer (no)
4 type (no)
5 do crosswords (yes)
6 swim (yes)

He can't play the piano.

6 Choose four or five places from the list below which are near where you are now, or where you live.

chemist   baker   supermarket
cinema   department store   library
station   market   bank   car park

Write sentences saying where they are.

There's a chemist in rue de Rivoli, next to the bookshop.

7 Write directions to the places you describe in 6.

Go along the Champs Elysées and turn left.

8 Think of five people you know. Write sentences saying what they're doing at the moment.

Brigitte is going to work. Frank is taking the train to London.

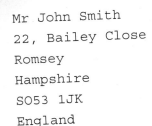

```
Mr John Smith
22, Bailey Close
Romsey
Hampshire
SO53 1JK
England
```

SOUNDS

1 ☐ Listen to these words. Is the underlined sound /ð/ or /θ/? As you listen, say the words aloud.

<u>th</u>is   <u>th</u>ink   <u>th</u>at   <u>th</u>ree   <u>th</u>e   mo<u>th</u>er

2 ☐ Listen to these words. Is the underlined sound /ð/ or /θ/?

<u>th</u>eir   <u>th</u>irty   <u>th</u>eatre   <u>th</u>ank

3 ☐ Listen to these words. As you listen, say the words aloud.

start	madam	mother
are	family	Monday
dark	map	lunch
park	have	uncomfortable

4 ☐ Listen to these words. Is the underlined sound /ɑː/, /æ/ or /ʌ/?

h<u>a</u>t   t<u>a</u>xi   <u>a</u>fternoon   r<u>u</u>n   b<u>a</u>th
f<u>a</u>st   c<u>a</u>r   b<u>u</u>s   m<u>a</u>rket   b<u>a</u>nk

5 ☐ Listen and underline the stressed words in these questions.

1 How do you get to work?
2 How long does it take?
3 Can you speak English?
4 Do you always cook dinner?

Now say the questions aloud. Make sure you stress the correct words.

WRITING AND SPEAKING

1 Look at the envelope above and find the following features.

title   first name   family name
county   number of home
road or street   town   country
post code

2 Write the following addresses in a similar way to the address on the envelope in 1.

1 mrs hillary jones 34 denver street oxford ox4 1sd
2 dr michael carey 4 horseferry road london e4 2sf
3 mr kenneth green 45 golden hill brighton sussex bn1 4fg england

3 Work in pairs. Do you write addresses on envelopes in a similar way? What information do you put first?

4 Go round the class asking for addresses of other students. Spell any difficult words.

16 | *Who was your first friend?*

Past simple (1): *be*

SPEAKING AND LISTENING

1 Think about when you were a child. Look at these questions and think about your answers.

☐ Who was your first friend?
☐ Where was your first school?
☐ What was the name of your first teacher?
☐ What was on the walls of your first classroom?
☐ What was your best birthday?
☐ What was the best day of the week?

2 🔲 Listen to four people talking about when they were a child. Put the number of the speaker by the question they answer in 1.

3 Work in pairs. Check your answers to 2.

4 Write your answers to these questions. *When you were a child ...*

– what was your favourite toy?
– what was your favourite game?
– what was your favourite food?
– what was your favourite drink?

5 Your teacher will ask some students to read out their answers to 4. As you listen, tick (✓) any words you hear which are the same as the words you wrote. How many students have the same answers?

GRAMMAR

a'farmativ

> **Past simple (1): *be***
>
> | Affirmative | Negative |
> |---|---|
> | *I was* | *I wasn't* |
> | *You were* | *You weren't* |
> | *He/she/it was* | *He/she/it wasn't* |
> | *We were* | *We weren't* |
> | *They were* | *They weren't* |
>
> **Questions**
> Who **was** your first friend?
> His name **was** John.
> **Were** you a happy child?
> Yes, I **was**. No, I **wasn't**.

1 Look at the grammar box. How many forms does the past simple of the verb *be* have?

2 Work in pairs. Ask and answer the questions in *Speaking and listening* activity 1 and 4.

Who was your first friend?
Her name was Sophia.

3 Write your partner's answers to the questions in 1.

George's first friend was Sophia.

SOUNDS

Listen to these sentences. Do you hear /wɒz/ or /wəz/?

1 My best friend was Jack.
2 Were you born in Rome? Yes, I was.
3 I was quite naughty.
4 My favourite drink was Coke.
5 What was your best birthday?
6 Were you happy? Yes, I was.

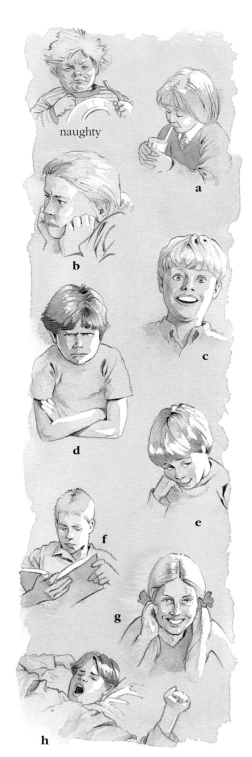

naughty

a

b

c

d

e

f

g

h

VOCABULARY AND SPEAKING

1 Look at these adjectives to describe character. Match them with the pictures.

naughty happy bad-tempered shy
stubborn lazy friendly serious
well-behaved

2 Which is the odd word out?

1 happy friendly naughty
2 bad-tempered happy lazy
3 well-behaved lazy stubborn

3 What were you like as a child? Choose adjectives to describe yourself.

Work in pairs. Now ask and say what you were like as a child.

What were you like as a child?
I was quiet, friendly, polite and well-behaved. What were you like?
I was naughty.

4 Work in pairs. Look at the photos of famous people. What do you think they were like as children?

I think the Queen was well-behaved as a child.
Yes, and I think she was serious.

5 Work in groups of four or five. Make a list of six famous people. Write their names on separate pieces of paper and fold them.

6 Choose one of the six pieces of paper. Make a list of adjectives to describe what you think the person on your piece of paper was like as a child.

7 Tell the others what you think the person in 6 was like as a child. Don't say who it is!

I think he was naughty and lazy ...

Can they guess who he/she is?

WRITING

1 Write three true statements and one false statement about you and your family when you were a child.

I was lazy at school.
I was happy at home.
My brother was very naughty.
My teacher's name was Marlon.

2 Work in pairs. Show your partner your statements. Write a question about any of your partner's statements.

My teacher's name was Marlon.
What was your teacher's family name?

3 Write answers to your partner's questions.

His surname was Brando.

4 Continue writing questions about your partner's statements until you guess the false one.

17 | *How about some oranges?*

Some and *any* (2); countable and uncountable nouns;
making suggestions

VOCABULARY AND LISTENING

1 Match the words in the box with the photo. Which items can't you see?

> apple bacon banana beef bread butter carrot
> cheese chicken coffee egg grapes lamb lettuce
> lemon oil onion orange juice potato rice tea
> tomato water peach cucumber

2 Put the items under these headings:
meat, fruit, vegetables, dairy products, drink.
Which three items are difficult to put under headings?

3 Work in pairs and check your lists in 2. Which items do you often eat or drink?

4 Decide where the sentences a – d go in the conversation.

JEAN OK, what do we need?
TONY We need some fruit and vegetables.
JEAN How about some oranges?
TONY (1) _____
JEAN Yes, there aren't any bananas. And let's get some apples.
TONY OK, apples. And we haven't got any onions.
JEAN (2) _____
TONY That's right, we haven't got any carrots. And let's get some meat.
JEAN Yes, OK. You like chicken, don't you?
TONY (3) _____
JEAN OK two kilos of tomatoes. Anything else?
TONY (4) _____
JEAN No, we need a couple of litres of water and let's get some juice. That's it.

a A kilo of onions. That's enough. And some carrots.
b No. Oh, have we got any water?
c Yes, chicken's great. And we need some tomatoes.
d OK, and we'll have some bananas.

🔊 Now listen and check.

GRAMMAR

> **Some and any (2)**
> You use *some* in affirmative sentences with uncountable nouns and plural nouns.
> *We need **some** fruit and vegetables.*
> You use *some* in questions when you ask for, offer or suggest something.
> *How about **some** oranges?*
> You use *any* in questions and negative sentences with uncountable nouns and plural nouns.
> *Have we got **any** water? Have we got **any** oranges?*
>
> **Countable and uncountable nouns**
> **Countable nouns have both a singular and a plural form.**
> *a banana – some bananas, a tomato – some tomatoes*
> **Uncountable nouns do not usually have a plural form.**
> *water, juice, coffee*
>
> **Making suggestions**
> ***Let's** have some apples.   **How about** some meat?*

1 Complete the sentences with *some* or *any*.

1 I need *some* water.
2 Have you got *any* onions?
3 I need to do *some* shopping.
4 We need *some* fruit and vegetables.
5 They haven't got *any* potatoes.
6 He wants *some* grapes.
7 How about *some* coffee?
8 There isn't *any* rice.

2 Make a list of everything you ate and drank yesterday. You can use a dictionary if you like.

toast, jam, coffee

Now work in pairs. Ask and say what you ate and drank.

I had some toast and jam, and two cups of coffee.
Did you have any cheese?
Yes, I did.

3 Look at the words in the vocabulary box. Write *C* for countable or *U* for uncountable.

apple C, ...

4 Write *a* or *some*.

1 book   2 money   3 tea   4 fruit
5 meat   6 pen   7 sandwich
8 chair   9 newspaper   10 coffee

5 Work in pairs and act out the conversation in *Vocabulary and listening*, activity 4.

6 Work in pairs. Make a list of the shopping you need to buy.

Now act out the conversation again with your own shopping list.

SOUNDS

1 [▢] Listen and repeat these words. Is the underlined sound /s/ or /z/?

apple<u>s</u>   baná́na<u>s</u>   carrot<u>s</u>
egg<u>s</u>   grape<u>s</u>   onion<u>s</u>   potatoe<u>s</u>
tomatoe<u>s</u>

2 [▢] Listen and repeat these phrases. Notice how *and* is pronounced /ən/.

fish and chips
salt and pepper
bread and butter
apples and pears
oranges and lemons
milk and sugar
meat and vegetables

LISTENING AND SPEAKING

1 Read these statements about food and drink. Say if they're true or false for your country.

	Your country	**Britain**
We eat eggs and bacon for breakfast.		
There's always meat and vegetables at the main meal.		
We always drink wine at lunch and dinner.		
We drink tea during the day.		
We often eat potatoes with our main meal.		
Many people don't eat meat.		

2 [▢] Listen to Lisa and find out if the statements in 1 are true or false for Britain. Complete the chart as you listen.

таблица .

3 Work in pairs and check your answers.

[▢] Now listen again and check.

18 *I was born in England*

Past simple (2): regular verbs; *have*

VOCABULARY AND READING

1 Tick (✓) the verbs you can use to describe events in your life.

> live learn work decide play
> start receive appear marry

Kazaτъcл

2 You're going to read about the life of the musician and singer Sting. Which of these words do you expect to see?

hit record teenager
rhythm and blues album
song film group award
politician teacher

Sting was born in Wallsend, England in 1952, and lived there for over 20 years. As a child he learned to play the piano and guitar. He worked as a teacher in Newcastle but then in 1978 he decided to create the group *The Police* with Stewart Copeland and Andy Summers. They had hits with *Message in a bottle* in 1979 and *Every breath you take* in 1983. *The Police* played together until 1984. Then Sting started to sing on his own and received awards for several songs. He appeared in the films *Brimstone and Treacle* in 1982, *Dune* in 1984 and *Plenty* in 1985. He married Trudy Styler in 1986.

3 Now read the passage and see if you guessed correctly in 2.

4 Read the passage and match the two parts of the sentences.

1 He was born in Wallsend
2 He learned to play the piano and guitar
3 He started work as a teacher
4 They finished playing together in 1984
5 He received awards for several songs

a but then created *The Police* in 1978.
b when he was a child.
c and worked as a teacher in Newcastle.
d then he appeared in three films.
e then Sting started to sing on his own.

GRAMMAR

> ### Past simple (2): regular verbs
>
> **You use the past simple to talk about an action or event in the past which is finished.**
> *They **finished** playing together in 1984.*
>
> **You form the past simple of most regular verbs by adding *-ed* to the infinitive.**
> *learn – learned   work – worked*
>
> **You add *-d* to verbs ending in *-e.***
> *live – lived*
>
> **The form is the same for all persons.**
>
> | *I* | | |
> |---|---|---|
> | *you* | | |
> | *he/she/it* | *lived* | *in London.* |
> | *we* | | |
> | *they* | | |
>
> **(For information about other regular endings, see Grammar review, page 108.)**
>
> ### Have
>
> **The past simple of *have* is *had*. The form is the same for all persons.**
> *They **had** hits with Message in a bottle.*

1 Read the passage again and find the past simple tense of the verbs in the box in *Vocabulary and reading* activity 1.

live – lived

2 What is the past simple tense of the following regular verbs?

die stay look like talk visit want finish open close watch

3 Complete the sentences with suitable verbs from 2.

1 I _____ the USA in 1996.
2 He _____ the football match on television in the evening.
3 He _____ to her for a long time.
4 The film _____ at half past ten.
5 Shakespeare _____ in 1616.
6 I _____ *Message in a bottle* very much.

SOUNDS

1 🔲 Listen to the *-ed* endings of these past simple verbs. Notice how /ɪd/ adds another syllable to the verbs.

/t/	/d/	/ɪd/
worked	lived	decided

2 Put these verbs in the correct columns above.

started learned visited closed died finished wanted stayed watched talked

🔲 Now listen and check. As you listen, say the words aloud.

LISTENING AND SPEAKING

1 Work in pairs. You're going to listen to a passage about the singer Whitney Houston.

Student A: Turn to Communication activity 3 on page 99.
Student B: Turn to Communication activity 23 on page 104.

WHITNEY HOUSTON stars as Rachel Marron in Warner Bros.' romantic thriller THE BODYGUARD. Also starring KEVIN COSTNER, THE BODYGUARD is a Tig Production produced by Lawrence Kasdan, Jim Wilson and Kevin Costner, and directed by Mick Jackson. Released in the UK by Warner Bros.

2 Work together and complete the chart with as much detail as possible.

> **Whitney Houston**
>
> Born: ....
>
> Started singing: ....
>
> First hit: ....
>
> Appeared in *The Bodyguard*: ....
>
> Number of copies sold of first two albums: ....
>
> *I will always love you*: ....

3 Match the two parts of the sentences.

1 She was 11 years old
2 She was 22 years old
3 She appeared in *The Bodyguard*
4 She worked with other singers
5 She received a Grammy Award in 1985

a when she had her first hit album.
b then in the following year she had a hit with a song from the film.
c before she had a hit with her first album.
d when she started singing.
e and in the next year was the first pop singer to sell 10 million copies.

4 Work in pairs. Think of a famous singer or musician in your country. You can use an encyclopaedia. Talk about:

– where he/she was born
– when he/she was born
– where he/she lived as a child
– how long he/she lived there
– when he/she started his/her career
– what he/she did and when

43

19 | *What's he like?*

Describing people (1)

VOCABULARY

1 Work in pairs. Match the words in the box with the photos.

> curly straight elderly
> young long short
> tall *[handwritten: высокий]*

2 Put the following adjectives under the following headings: *height, size, hair, age, general impression.* *[handwritten: впечат.]*

> attractive curly dark
> elderly fair straight *[handwritten: справедл.]*
> good-looking long
> medium height young
> well-built small tall
> middle-aged old
> slim pretty short

> *height:*
> *medium-height . . .*

3 You can use the nouns in the box to describe physical features. *[handwritten: физич. черты]*

> beard glasses hair
> moustache eyes *[handwritten: усы]*

[handwritten left margin: bi(ə)rd борода]

Look at the photos. Find someone who has: *[handwritten: бор.]*

[handwritten: светлые вол.]

– short hair   – fair hair
– a beard   – glasses
– a moustache

FUNCTIONS

> **Describing people (1)**
> **You use the following expressions to describe people.**
> *What's he like?*
> *He's middle-aged and good-looking.*
> *What does he look like?*
> *He's quite tall, with dark hair.*
> *He looks like his mother.*
> *He's got a moustache. She's got glasses.*
> *How tall is he?*
> *He's one metre seventy-five.*
> *How old is he?*
> *He's twenty.*
> **You can also use *What's he/she like?* to ask about someone's character.** *[handwritten: ˈkærɪktər]*
> *What's she like?*   *She's very nice.*

1 🔊 Look at the photos. Listen and decide who the speaker is describing.

2 Choose one of the people in the photos. Write words to describe him or her.

Now work in pairs. Show your partner the list of words. Can you guess who your partner is describing?

3 Work in pairs. Describe the following people to each other.

– one of your parents or grandparents
– a brother or sister
– your best friend
– one of your parents' friends
– someone in the class

WRITING

1 Read the letter. Does Nick know Pat?

> 33, Highclere Road
> Basingstoke
> England
> 20 May
>
> Dear Pat,
>
> Thanks for your letter. It's very kind of you to let me stay with you and to meet me at the airport.
>
> I arrive at 22.30 on Thursday 22 June. I'm twenty and I've got dark curly hair. I'm quite short, about one metre seventy and I'm quite well-built. If you have time, send me a description of yourself so I can recognise you.
>
> See you at the airport.
>
> Best wishes,
>
> Nick

2 Decide which of the people in the photos is Nick.

3 Write a letter to Nick describing yourself. Describe:

– your age   – your hair   – any physical features

– your height   – your size

Make sure you put your address and the date on the letter.

LISTENING AND SPEAKING

1 Look at these statements and decide if they are mostly true or false for your country.

1 People are tall when they're over one metre sixty centimetres.
2 People are old when they're over sixty.
3 People are middle-aged when they're over forty.
4 People are usually quite well-built.

2 🔲 Listen to Kevin, who is Irish, talking about the statements. Are they true or false for his country?

3 Work in pairs. What did Kevin say about the statements?

1 People are tall when they're over:
   a one metre sixty centimetres
   b one metre eighty centimetres c two metres
2 People are old when they're:
   a fifty b sixty c seventy
3 People are middle-aged when they are:
   a forty b fifty c sixty
4 People are usually:
   a well-built b slim c well-built or slim

Can you remember any other details?

🔲 Now listen again and check.

20 A grand tour

Past simple (3): irregular verbs; *yes/no* questions and short answers

SOUNDS

1 Here are the names of some cities in Europe. Match them with the English name of the country they're in.

> Budapest   Zurich   London   Paris
> Venice   Vienna

> Austria   England   France   Hungary
> Italy   Switzerland

2 🔲 Listen and check your answers to 1. As you listen, say the words aloud.

3 What's the English name for your country or city? Is it the same as in your language?

VOCABULARY AND LISTENING

1 Match the verbs in the box with the phrases below.

> buy   do   find   fly   go   have
> listen   lose   make   read   stay
> take   visit   watch   write

root = маршрут

to the Opera	some souvenirs
a museum	a meal
some shopping	a cheap hotel
your wallet	home
to a concert	friends
the newspaper	some postcards
with friends	a tram
the football	

2 Work in pairs and look at the map of Europe. Find two or three places you want to visit. Can you think of things you want to do there?

3 🔲 Listen to Mary and Bill from the USA who went on a tour around Europe. Follow their route on the map.

root – маршрут.

4 Before Mary and Bill started their tour, they made a list of things they wanted to do. Tick (✓) the things they did.

пробовал баню

buy some souvenirs	☐	have a steam bath ☐
visit the museums and galleries	☐	write some postcards ☐
fly home	☐	go to the Opera ☐
do some shopping	☐	take a tram ☐
find a cheap hotel	☐	lose a wallet ☐
stay with friends	☐	have a meal in a smart restaurant ☐
meet some New Yorkers	☐	read the English newspapers ☐
make friends with local people	☐	relax in the parks ☐
listen to a concert	☐	see the Queen ☐

🔲 Listen again and check.

GRAMMAR

Past simple (3): irregular verbs

Many verbs have an irregular past simple form.

buy – bought   write – wrote
go – went

For a list of irregular verbs, see page 114.

Yes/no questions and short answers

You form *yes/no* questions with *did*.

Did they stay with friends in Paris?
Yes, they did. No, they didn't.

---

1 Match the infinitives with their irregular past tenses.

become buy come find fly have hear get give go leave lose make read say sell send spend wake win write

had came left went became wrote got gave won lost found bought sold read spent woke said flew made sent heard

2 Complete the sentences with *did, was* or *were*.

1 _____ they go to the Eiffel Tower?
2 _____ they born in the USA?
3 _____ they have a good time?
4 _____ it sunny in France?
5 _____ they do some shopping in Paris?
6 _____ they make friends with anyone?

3 Work in pairs. Look at your answers to *Vocabulary and listening* activities 3 and 4 and say where Mary and Bill went and what they did.

They went to Paris and did some shopping.

4 Give short answers to the questions.

1 Did Bill have a steam bath in Budapest?
2 Did they see the Queen in London?
3 Did they watch the football in Paris?
4 Did they take a tram in Vienna?
5 Did they find a cheap hotel in Vienna?
6 Did Bill lose his wallet in Budapest?
7 Did they make friends in Venice?
8 Did they write some postcards in Budapest?

1 Yes, he did.

SPEAKING

1 Ask other students about their last holiday. Find someone who:

took a plane
stayed with friends
did some sightseeing
lost their passport
stayed in a hotel
went to Britain
had a meal in a restaurant
stayed in a tent
made some new friends

Did you take a plane?
Yes, I did.
Did you stay with friends?
No, I didn't.

2 Work in pairs. Tell each other about a holiday you had. Talk about:

– where you went
– what you did
– where you stayed
– how long you stayed there
– what the weather was like
– when you went home

Progress check **16–20**

VOCABULARY

1 When you write down new words, write down other words which you associate with them. They don't have to be the same part of speech.

musician – *song, guitar, rhythm and blues*
fly – *plane, airport*

Match the words in box A with the associated words in box B. There may be more than one possibility.

A | hit record bread polite marriage attractive restaurant friendly |

B | album well-behaved divorced cheerful butter good-looking meal |

hit record – album

2 You can use word charts to write down your new words.

TRANSPORT			
Road	**Rail**	**Air**	**Sea**
bicycle, car, bus garage, bus stop	train, underground station	plane airport	ship, boat ferry, port

Write word charts for one of the following. Use a dictionary, if necessary.

Home – (think of rooms and furniture)
Time – (think of days and months)
Places – (think of adjectives to describe a town, facilities etc)

3 It's also useful to write down the other parts of speech of a word.

Noun	**Verb**
writer	write
success	succeed

Use your dictionary to find:
– nouns formed from the following verbs

decide play marry visit start

– nouns formed from the following adjectives

happy lazy attractive young

GRAMMAR

1 Complete the sentences with *some* or *any*.

1 I'd like _____ apples, please.
2 Do we need to get _____ rice?
3 They want _____ tea.
4 I haven't got _____ money on me.
5 Let's buy _____ apples.
6 There isn't _____ toast.

2 Write C for countable or *U* for uncountable.

water carrot juice egg onion
lamb apple milk oil burger
cheese pasta potato

water U, ...

3 Write the past simple form of the following verbs.

become work decide buy
appear visit have go die start
lose land paint read invent
send discover

4 Match the questions 1 – 4 with the answers a – d.

1 What's she like?
2 What does she look like?
3 How tall is she?
4 How old is she?

a About one metre seventy.
b She's tall with long, dark hair.
c About twenty.
d She's very pretty.

5 Write a description of the people in the picture.

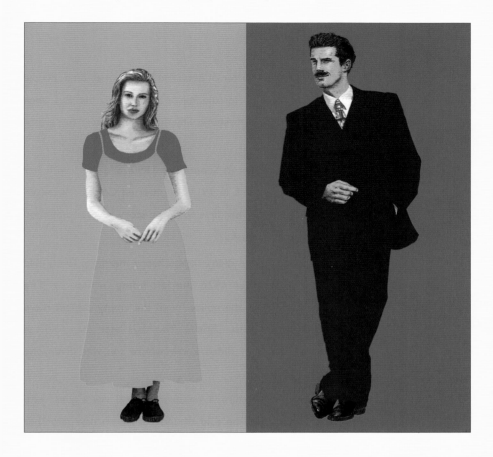

6 Complete the questions with *did, was* or *were*.

1 _____ Jonas born in Vienna?
2 _____ there anything to eat at the party?
3 _____ you see the football match last night?
4 _____ there lots of people on the bus?
5 _____ they arrive on time?
6 _____ she get home safely?

7 Here are the answers to the questions in 6. Write short answers.

1 No 2 Yes 3 Yes 4 No 5 No 6 Yes

1 No, he wasn't.

SOUNDS

1 Group the words which have the same underlined sound.

aw<u>ar</u>d b<u>ou</u>ght div<u>or</u>ced g<u>o</u>t l<u>ear</u>ned l<u>o</u>st m<u>or</u>ning ret<u>ur</u>n s<u>er</u>ved th<u>ir</u>d w<u>a</u>s w<u>or</u>d

How many groups are there?

[cassette icon] Now listen and check. As you listen, say the words aloud.

2 Many words have letters which you don't pronounce.

~~eight~~ wo~~r~~d lis~~t~~en

Cross out the silent letters in these words.

card daughter cupboard bought island lamb

[cassette icon] Now listen and check. As you listen, say the words aloud.

3 Match the words with the stress patterns.

☐☐   ☐☐☐   ☐☐☐

award appear teacher receive unhappy eleventh evening tomato banana

[cassette icon] Listen and check. As you listen, say the words aloud.

4 [cassette icon] Listen to the questions in *Grammar* activity 6. Put a tick (✓) if you think the speaker sounds interested.

Now say the questions aloud. Try to sound interested.

SPEAKING

Work in groups of three or four. Play *Twenty Questions*.

Student A: Think of a person. Answer Student, B, C and D's questions with *Yes* and *No* until they find out who the person is.

Student B, C and D: Student A is thinking of a person. In turn, ask questions to find out who the person is. You can only ask twenty questions in all.

1 Is it a man?   Yes.
2 Is he alive?   No.
3 Was he a singer?   Yes.

Change round when you're ready. *супец 6*

Past simple (4): negatives; *wh-* questions

Agatha Christie was the most successful writer of detective stories of all time. People all over the world read her stories of Hercule Poirot and Miss Marple. But when she died in 1976 there was a final mystery: why did she disappear for eleven days in December 1926?

Agatha Christie was born in September 1890. She lived with her family in Devon, England. In 1914 she married Colonel Archibald Christie. She wrote her first detective story in 1920 and soon she was very successful.

But Agatha Christie didn't have a happy marriage. On a cold night in December 1926 she left home in her car. The following morning, the police found the empty car but there was no sign of Agatha Christie. Two days later, they told the newspapers that they didn't know where she was. Everyone thought she was dead.

But 250 miles away in Yorkshire, a waiter in a hotel saw a guest who looked like Agatha Christie, and he told the police. Eleven days after her disappearance, her husband found her again in the hotel dining room.

The couple were soon divorced. She married Sir Max Mallowan, an archaeologist, in 1930 and she continued to write her mysteries. But she didn't explain what happened in 1926. Did she want to kill herself? Did she want to show her husband that she didn't love him? Did she hope to sell more books?

Over the years, Agatha Christie wrote more than 80 mysteries and sold over 300 million books. But she didn't tell anyone why she disappeared in December 1926.

PENGUIN BOOKS

MYSTERY AND CRIME

MURDER ON THE ORIENT EXPRESS

AGATHA CHRISTIE

MYSTERY AND CRIME

Why did Agatha Christie disappear?

VOCABULARY AND READING

1 Work in pairs. You're going to read an article about Agatha Christie. Do you know who she was? Talk about anything you know about her. Think about what you'd like to find out about her.

2 Work in pairs. Look at the title of the article and the words in the box. What do you think happened?

> successful writer detective story
> mystery marriage unhappy
> left home disappear waiter hotel
> guest husband find divorced
> kill tell

3 Read the article and choose the correct answer to the question in the title.

1 Because she was unhappy.
2 We don't know.
3 Because she wanted to kill herself.

Did you guess correctly in 2?

4 Put a tick (✓) by the statements which are true.

1 Agatha Christie married when she was fifteen.
2 The police found her in the car.
3 The police said that she was dead.
4 A hotel waiter recognised her.
5 Her husband found her at home.
6 She married a crime writer.
7 She wrote over eighty detective stories.
8 She died at the age of seventy-six.

5 Work in pairs.

Student A: Turn to Communication activity 8 on page 100.

Student B: Turn to Communication activity 15 on page 102.

GRAMMAR

> Past simple (4): negatives ~~кроме~~
> **You form the negative for all verbs except** *be* **with** *didn't* **+ infinitive without** *to.*
> She **didn't have** a happy marriage. She **didn't explain** her disappearance.
>
> *Wh-* questions
> **You form** *Wh-* **questions with a** *Wh-* **question word** *(who, what, when, why)*
> **+** *did* **+ infinitive without** *to.*
> **What did** Agatha Christie **do?** **Why did** she **disappear?**
> Where **was** she born?

1 Correct the false statements in *Vocabulary and reading* activity 4.

She didn't marry when she was fifteen. She married when she was twenty-four.

2 Here are some answers about Agatha Christie. Write the questions.

1 She wrote detective stories.
2 In September 1890.
3 In Devon.
4 Colonel Archibald Christie.
5 In December 1926.
6 In a hotel in Yorkshire.
7 Eleven days after her disappearance.
8 He was an archaeologist.

3 Turn to Communication activity 6 on page 100.

SOUNDS

Notice how you don't usually stress auxiliary verbs.

[cassette icon] Listen to the pronunciation of the auxiliary verbs in the questions you wrote in *Grammar* activity 2.

Now say the sentences aloud.

WRITING AND SPEAKING

1 Write a short autobiography. Say:

– where you were born – when you started school
– where you lived – if there were any special events in your life

I was born in 1982. I lived with my family in Malaga. Then we moved to Seville.

2 Work in pairs and exchange your autobiographies. Write extra questions about your partner's autobiography.

How long did you live in Malaga? When did you move to Seville?

3 Read your partner's questions. Rewrite your autobiography and include the answers.

I was born in 1982 in Malaga, and I lived there with my family for ten years. Then we moved to Seville in …

22 | Dates

Past simple (5); expressions of time

VOCABULARY AND SOUNDS

1 Match the words in the box with the numbers below.

> eighth   eleven   fifth   first   fourth   ninth   second
> seventh   sixth   tenth   third   twelfth

1st   2nd   3rd   4th   5th   6th   7th   8th   9th   10th
11th   12th

1st – first

2 Write the words for the following numbers.

13th   17th   20th   21st   22nd   23rd   27th   30th   31st

13th – thirteenth

3 Notice how:

– you write
1st March   13th April

– you say
the first of March   the thirteenth of April

[cassette] Now listen and repeat these dates.

1st March   13th April   23rd September
4th January   30th July   2nd May   20th June
10th October

4 Work in pairs. What dates are the following?

New Year's Day   your birthday
the national day or an important day in
your country   yesterday   today

| January   February   March   April   May   June July |
| August   September   October November   December |

New Year's Day is the first of January.

LISTENING AND SPEAKING

1 Match these words with the special days below.

present   church   letter   party   reception
driving licence   certificate   card   forget   ring

- [ ] an important birthday   - [ ] a wedding day
- [ ] passing an exam   - [ ] an anniversary
- [ ] Independence Day

2 🔲 You're going to hear three people talking about one of the special days. Listen and put the number of the speaker by the special day he/she is describing in 1.

3 Work in pairs and check your answers to 2. Which speaker uses the following expressions?

last Thursday   yesterday evening   five months ago
in 1987   in August   from nine to five
on the eleventh of December   at the end of the year

🔲 Listen again and check.

GRAMMAR

Past simple (5): expressions of time
You can use these expressions of past time to say when something happened.

last   *night/Thursday/August/month/year*
*I saw him **last year.***

in   *August/1987*
*We got married **in 1987.***

ago   *days/weeks/months ago*
*We went on holiday **three weeks ago.***

yesterday   *morning/afternoon/evening*
*I met her **yesterday morning.***

at the end of   *the day/month/year*
*My birthday is **at the end of the month.***

on   *Monday/the eleventh of December*
*My birthday is **on the eleventh of December.***

1 Complete the sentences with an expression of time in the grammar box.

1 I went to the dentist _____.
2 I did my homework _____.
3 I started learning English _____.
4 I bought someone a present _____.
5 I met my best friend _____.
6 I wrote a postcard _____.

2 Work in pairs. Ask and answer the questions.

1 When did you last buy a new coat?
2 When did you get home yesterday?
3 When did you last go to the cinema?
4 When did you get up this morning?
5 When did you last have a birthday?
6 When did you start this lesson?

1 A year ago.

SPEAKING

1 Work in pairs and answer the questions in the quiz. You score 2 points if you get the right answer, and 1 point if it's very close. Your teacher will decide.

When did ...

1 ... the Berlin Wall come down?
2 ... the Russian Revolution start?
3 ... astronauts first land on the moon?
4 ... the first atom bomb explode?
5 ... Columbus discover America?
6 ... the Chinese build the Great Wall?
7 ... the Second World War start?
8 ... John Kennedy die?
9 ... the First World War start?
10 ... you start learning English?

2 Work in two pairs and continue the quiz in 1.

Pair A: Turn to Communication activity 5 on page 100.
Pair B: Turn to Communication activity 22 on page 104.

23 | *What's she wearing?*

Describing people (2); present continuous or present simple

VOCABULARY AND LISTENING

1 Look at the people in the photos and say what they're wearing. Use the words in the box to help you.

> trousers jeans skirt socks shorts dress tie shirt T-shirt jacket sweater shoes trainers

брюки
трайзерс
sHərt
кросовки
Swétər

Turn to Communication activity 13 on page 102 to find out what the other words are.

2 Look at the words in the box again. Find things:

– you wear on your feet   – you wear when it's hot
– that men usually wear   – you wear when it's cold
– that women usually wear

3 Which words can you use to describe the clothes:

– you're wearing at the moment?   – you usually wear?

> comfortable fashionable casual smart warm

модный
повседневный
случайный

4 Look at the photos again and use these words to describe the people.

> sit down stand smile laugh wear

5 🔲 Listen to Jan and find out who the people in the photos are.

6 Work in pairs and check your answers. Complete the chart with as much detail as possible.

🔲 Listen again and check.

	Harriet	**John**	**Edward**	**Louise**
What's he/she wearing?				
What's he/she doing?				

FUNCTIONS AND GRAMMAR

> ### Describing people (2)
>
> *She's wearing a black dress.*   *She's sitting down.*
> *She's smiling.*
> *He's the man sitting down in the armchair.*
>
> ### Present continuous or present simple
> **You use the present continuous to say what is happening now or around now.**
> *She's **wearing** a skirt.*   *He's **living** in Oxford at the moment.*
>
> **You use the present simple to describe something which is true for a long time.**
> *She usually **wears** jeans.*   *He **lives** in London.*

1 Choose the correct verb form.

1 He usually *wears/is wearing* a suit.
2 She *smiles/is smiling* at him.
3 Guy *is standing/stands* by the door.
4 Maria is the person who *is talking/talks* to Ken.
5 Where *are you working/do you work* at the moment?
6 You usually *stand up/are standing up* when you meet someone.
7 Ricardo *smokes/is smoking*, but he *doesn't smoke/isn't smoking* at the moment.
8 Pierre *wears/is wearing* brown shoes every day.

2 Work in pairs.

Student A: Choose someone in the class and describe him/her. Don't say who he/she is.

Student B: Listen to Student A's description of someone in the class. Can you guess who it is?

He's wearing jeans.
He's sitting down.

3 Write full answers to the questions in the chart in *Vocabulary and listening* activity 6.

Harriet's wearing jeans. She's standing by the door.

4 Work in pairs.

Student A: Turn to Communication activity 7 on page 100.

Student B: Turn to Communication activity 27 on page 105.

READING AND SPEAKING

1 Read the questionnaire and answer the questions.

> # What do your clothes say about you?
>
> **1** You see someone with blue hair wearing a yellow jacket and red trousers. What do you do?
> a smile   b laugh   c wear the same clothes
>
> **2** You are going to an interview. What do you wear?
> a jeans   b a suit   c something comfortable
>
> **3** You're going to work. What do you wear?
> a trousers   b trainers   c a jacket
>
> **4** You're going to a party. What do you wear?
> a a jacket   b a T-shirt   c a suit/dress
>
> **5** You're buying a new jacket. What colour do you buy?
> a black   b red   c orange
>
> **6** You're buying clothes for cold weather. Which is more important?
> a comfort   b warmth   c fashion
>
> **7** You want to give a good impression. Which style do you choose?
> a comfortable but smart   b smart and formal
> c casual
>
> **8** What kind of clothes do you prefer?
> a cheap   b expensive   c fashionable
>
> **9** You're going to play tennis with a friend. What do you do wear for the game?
> a a tie   b shorts
> c a sweater
>
> **10** It's very hot at work or school. What do you do?
> a wear shorts   b take off your jacket or sweater
> c do nothing

2 Work in pairs and talk about your answers to the questionnaire.

3 Turn to Communication activity 25 on page 105 and find out what your clothes say about you.

24 | *I'm going to save money*

Going to; because and so

READING AND LISTENING

1 Read the passage *My New Year's resolution*.
Who do you think you can see in the photos?

My New Year's resolution ...

1 'I'm going to see my friends more often.' *Phil*

2 'I'm going to save money.' *Harriet*

3 'I'm going to change my job.' *Pete*

4 'We're going to travel around Europe.'
Andrew and Mary

5 'We're going to have French lessons.' *Jill and Steve*

6 'I'm going to spend more time with my parents.' *Jenny*

7 'We're going to invite more friends for dinner.'
Henry and Celia

8 'I'm going to get fit.' *Kate*

9 'I'm not going to take work home.' *Dave*

10 'We're going to move.' *Judy and Frank*

2 Match the resolutions in 1 with the reasons below.

a We don't speak any foreign languages.

b We don't entertain very much.

c I hate my work.

d I stay at home all the time.

e We always stay in Britain for my holiday.

f Our house is too small.

g I never see my family.

h I want to spend more time with my children.

i I spend too much.

j I don't take enough exercise.

3 🔲 Listen to four people talking about their resolutions and the reasons. Find out who's speaking.
Did you guess correctly in 2?

4 Work in pairs. Can you remember any other details about what the speakers said?

🔲 Now listen again and check.

GRAMMAR

> *Going to*
>
> **You use *going to* + infinitive:**
> **– to talk about future intentions or plans.**
> *I'm **going to** see my friends more often.*
> *I'm **not going to** take work home.*
> **– to talk about something which we can see now is**
> **sure to happen in the future.**
> *I'm **going to** have a baby.*
>
> *Because* and *so*
> **You can join two sentences with *because* to describe**
> **a reason.**
> *Judy and Frank are going to move **because** their house is*
> *too small.*
>
> **You can join the same two sentences with *so* to**
> **describe a consequence.** kānsikwansnoeuegīlue
> *Judy and Frank's house is too small **so** they're going to move.*

1 Look at these sentences and explain the difference
between them.

I go to work at nine o'clock.
I'm going to go to work at nine o'clock.

2 Work in pairs. Check your answers to *Reading and
listening* activity 3. Say what the speakers are
going to/not going to do.

3 Work in pairs. Say what you're going to do this
weekend. Here are some ideas:

get up late   do some housework   play football
have a meal out   read the newspapers
meet some friends   watch television   go for a walk

4 Choose three sentences in *Reading and listening*
activities 1 and 2 and join them with *because*.

*Harriet is going to save money because she spends
too much.*

5 Choose three more sentences in *Reading and listening*
activities 1 and 2 and join them with *so*.

*Kate doesn't take enough exercise, so she's going to
get fit.*

VOCABULARY AND WRITING

1 Here are some verbs from this lesson. Can you
remember which nouns or noun-phrases they went
with in *My New Year's resolution*?

save   take   spend   get   invite   change

save money   ...

2 Which phrases in 1 can you use to describe what
you're going to do this weekend?

3 Write what you're going to do before you finish
Reward Elementary.

I'm going to save more money.

4 Write why you're going to do the things you
wrote in 3.

I want to go on holiday.

5 Join the sentences you wrote in 3 with the sentences
you wrote in 4 using *because*.

*I'm going to save more money because I want to go
on holiday.*

6 Make a class collection of resolutions and keep them
safe. When you finish *Reward* Elementary, read them
out and find out if anyone has kept their resolutions.

25 | *Eating out*

***Would like*; talking about prices**

VOCABULARY AND LISTENING

1 Look at the words in the box. Which pictures of food and drink can you see on the menu?

> burger Coke French fries ice cream sandwich salad pizza
> apple pie cheesecake kebab risotto pasta juice steak
> chocolate mousse coffee mayonnaise strawberry

многизу

FREDDY'S FAST FOOD

Burger	£2.50	
French fries	£1.25	
Steak	£4.25	
Pizza	£3.15	small
	£3.95	large
Salad	£1.50	
Pasta	£1.75	small
	£2.25	large
Apple pie	75p	
Ice cream	£1.25	
Chocolate mousse	95p	
Coke	75p	regular
	95p	large
Coffee	£1.25	

2 Which is the odd word out?

1 burger Coke juice coffee
2 apple pie cheesecake ice cream pasta
3 pizza pasta steak risotto
4 French fries chocolate mousse salad mayonnaise

3 Look at the food items in the box. Say what you like.

4 Look at this conversation in a *fast food* restaurant. Decide where the sentences a – f go in the conversation.

ASSISTANT Good afternoon. Can I help you?
CUSTOMER (1) _____
ASSISTANT Would you like a regular or a large Coke?
CUSTOMER (2) _____.
ASSISTANT Would you like anything else?
CUSTOMER (3) _____ *вкус*
ASSISTANT What flavour would you like?
CUSTOMER (4) _____
ASSISTANT OK.
CUSTOMER (5) _____
ASSISTANT That's four pounds fifty.
CUSTOMER (6) _____
ASSISTANT Thank you.

a A regular, please.
b Good afternoon. Yes, I'd like a burger with fries and a Coke, please.
c Strawberry, please.
d How much is that?
e Yes, I'd like an ice cream, please.
f Here you are.

5 🔲 Listen and check your answers to 4.

FUNCTIONS

> *Would like*
>
> **You use *would like* + infinitive to ask for something politely.**
> *I'd like a burger and a Coke, please.*
> ***Would you like** a regular or a large Coke?*
> *I'd like an ice cream, please.*
> *What flavour **would you like?***
>
> **Remember**
> **– you use *like* to say what you like all the time.**
> *I **like** Coke.* (= always)
>
> **– you use *would like* to say what you want now.**
> *I'd **like** a Coke.* (= now)
>
> **Talking about prices**
> *How much is it?*   *Two pounds ninety-nine.*

1 Choose the correct answer.

1 Would you like a drink?
  a No, I don't like a drink.   b No, thank you.

2 What would you like to eat?
  a I'd like a pizza.   b I like pizza.

3 Would you like some coffee?
  a Yes, please.   b Yes, I like some wine.

4 Do you like pasta?
  a Yes, please.   b Yes.

5 Can I help you?
  a Yes.   b Yes, I'd like a burger.

6 Would you like to order?
  a Yes, a burger, please.   b Yes, I like burger.

разыграть

2 Work in pairs and act out the conversation in *Vocabulary and listening* activity 4. Choose other items from the menu.

3 Work in pairs. Look at the menu and ask and say how much things cost.

SOUNDS

🔊   Listen and tick (✓) the sentences you hear.

1 a I like beer.   b I'd like a beer.
2 a He likes ice cream.   b He'd like ice cream.
3 a We like the bill.   b We'd like the bill.
4 a I like pizza.   b I'd like pizza.
5 a She'd like a salad.   b She likes a salad.
6 a Would you like pasta?   b Do you like pasta?

READING AND SPEAKING

1 You're going to read a passage about eating out in the USA. First, check you know what the underlined words in the passage mean.

2 Read *Eating out in the USA* and find out what it says about:

  – types of restaurants   – where to sit   – who to pay
  – how much to tip   – other advice

EATING OUT IN THE USA

In the USA, there are many types of restaurant. Fast food restaurants are very famous, with McDonalds and Kentucky Fried Chicken in many countries around the world. You look at a <u>menu</u> above the <u>counter</u>, and say what you'd like to eat. You pay the person who serves you. You take your food and sit down or take it away. When you finish your meal, you put the <u>empty container</u> and paper in the <u>rubbish bin</u>. There's no need to leave a tip.

In a coffee shop you sit at the counter or at a table. You don't wait for the <u>waitress</u> to show you where to sit. She usually brings you coffee when you sit down. You tell her what you'd like to eat and she brings it to you. You pay the <u>cashier</u> as you leave. A diner is like a coffee shop but usually looks like a <u>railway carriage</u>.

In a family restaurant the atmosphere is casual, but the waitress shows you where to sit. Often the waitress tells you her name, but you don't need to tell her yours. If you don't eat everything, your waitress gives you a doggy bag to take your food home. You <u>add</u> an extra fifteen per cent to the <u>bill</u> as a tip.

In top class restaurants, you need a reservation and you need to arrive on time. The waiter shows you where to sit. If you have wine, he may ask you to <u>taste</u> it. You can only <u>refuse</u> it if it tastes bad, not if you don't like it. When you get your bill, check it and then add fifteen to twenty per cent to it as a tip. You pay the waiter.

3 Work in pairs and check your answers. Compare eating out in the USA with eating out in your country.

Progress check **21–25**

VOCABULARY

1 When you see a word you don't understand, stop and ask yourself these questions:

– what is its part of speech?
– can I guess what it means?
– can the rest of the passage help me understand it?

Try not to use a dictionary or ask your teacher every time. *бодер очень*

Look at this extract from the passage in Lesson 25. Some of the difficult words are missing. Think about your answers to the questions above. Don't try to remember the exact word. *изблечеuuu*

> In the USA, there are several types of restaurant. *Fast food* restaurants are very famous, with McDonalds and Kentucky Fried Chicken in many countries around the world. You look at a _____ above the _____, and say what you'd like to eat. You pay the person who serves you. You take your food and sit down or take it away. When you finish your meal, you put the _____ container and paper in the rubbish bin. There's no need to leave a tip.
>
> In a *coffee shop* you sit at the counter or at a table. You don't wait for the _____ to show you where to sit. She usually brings you coffee as soon as you sit down. You pay the _____ as you leave. A *diner* is like a coffee shop but usually looks like a railway carriage.

Now look back at page 59 to find out what the missing words are.

2 Find ten words in the word puzzle. They go in two directions (↓) and (→). Five words are to do with food and drink and five words are to do with clothes.

тема – предметы обсуж.

3 Choose one of the other topics in Lessons 21 to 25 and make a word puzzle.

When you're ready, work in pairs and do each other's word puzzles.

GRAMMAR

1 Complete the sentences with *ago, from, to, last, yesterday, during, at.*

1 He started work *yesterday* morning.
2 We went camping *during* the summer.
3 I started learning French *last* year.
4 He rang me five minutes *ago*.
5 It was open *from* nine *to* five.
6 He left *at* the end of the week.

60

2 Choose the correct verb form.

1 I *wear/am wearing* jeans most of the time.

2 He's the man who *stands/is standing* next to Jim.

3 He *smokes/is smoking* twenty cigarettes a day.

4 I always *shake/am shaking* hands when I meet someone.

5 At the moment she *smiles/is smiling* at him.

6 I *speak/am speaking* fluent English.

1 I wear jeans most of the time.

3 Write five things which you're going to do next month.

I'm going to visit my sister next month.

4 Join the two sentences by rewriting them with *because*.

1 I'm hungry. I'd like something to eat.

2 I'm afraid we haven't got any pizza. Would you like pasta?

3 I like him. I'm going to see Tom Cruise's new film.

4 I'm going to live in London. I like the people there.

5 He can't see. He hasn't got his glasses.

6 We went on holiday last year. We're not going away this year.

1 I'd like something to eat because I'm hungry.

5 Rewrite the sentences in 4 with *so*.

1 I'm hungry so I'd like something to eat.

SOUNDS

1 🔈 Listen and repeat the following words.

meet their know sea son write our buy eye right no sun for there by I four knows too hour see meat nose two

Write the pairs of words which sound the same but have different spelling.

2 Say these words aloud. Is the underlined sound /ʊ/ or /uː/?

f<u>oo</u>d sh<u>oe</u> y<u>ou</u> g<u>oo</u>d s<u>ou</u>p c<u>oo</u>k b<u>oo</u>k d<u>o</u> f<u>oo</u>t c<u>oo</u>l p<u>u</u>t j<u>ui</u>ce b<u>oo</u>t

🔈 Now listen and check. As you listen, say the words aloud.

3 Look at these words. Underline the stressed syllable.

banana cabbage potato bacon trousers trainers cashier risotto salad hamburger toothpaste

🔈 Listen and check. As you listen, say the words aloud.

4 🔈 Listen to these questions. Put a tick (✓) if you think the speaker sounds polite.

1 Can I help you?

2 What would you like to eat?

3 Would you like a drink?

4 Would you like anything else?

Now say the questions aloud. Try to sound polite.

WRITING AND SPEAKING

1 Work in groups of three or four. You're going to prepare a quiz about important historical facts.
Write at least ten questions about important events/people in history. Think about:

– life and death of famous people

– inventions and discoveries

– wars

– political events

– artistic creations

2 Work with another group. In turn, ask and answer questions from your quizzes. You score one point for each correct answer. The group with the most points is the winner.

26 | *Can I help you?*

Reflexive pronouns; saying what you want to buy; giving opinions; making decisions

SPEAKING AND LISTENING

1 Match the sentences 1 – 4 with the sentences a – d to make four conversations.

1 Can I carry that for you?
2 Did you make it?
3 Are you buying this for yourself?
4 They've got themselves a new one.

a No, it's for a friend.
b Have they? What kind?
c No, Geoff made it himself.
d No, it's OK. I can carry it myself.

2 Work in pairs and check your answers to 1. Choose one conversation and write two or three sentences before and after it. Act out your conversation to the rest of the class.

3 Look at this conversation. Decide if the customer is buying something:

– for herself   – for someone else

ASSISTANT   Can I help you?

CUSTOMER   Yes, I'm looking for a T-shirt.

ASSISTANT   We've got some T-shirts over here. What colour are you looking for?

CUSTOMER   This green one is nice.

ASSISTANT   Yes, it is. Is it for yourself?

CUSTOMER   Yes. Can I try it on?

ASSISTANT   Yes, go ahead.

CUSTOMER   No, it's too big. It doesn't fit me. Have you got one in a smaller size?

ASSISTANT   No, I'm afraid not. What about the yellow one?

CUSTOMER   No, I don't like the colour. Yellow doesn't suit me. OK, I'll leave it. Thank you.

ASSISTANT   Goodbye.

4 🔲 Listen and underline anything that is different from what you hear.

5 Work in pairs and correct the conversation.

GRAMMAR AND FUNCTIONS

> **Reflexive pronouns**
> **You usually use a reflexive pronoun when the subject and the object of a sentence are the same.**
> *myself yourself himself herself itself ourselves yourselves themselves*
> *I can carry it **myself.***
>
> **Saying what you want to buy**
> *I'd like a ...*
> *I'm looking for ...*
> *Can I try it on?*
> *Have you got a/any ...?*
> *Have you got it in another colour?*
>
> **Giving opinions**
> *It's too big/small/long/short.*
> *It doesn't suit me.*
> *It doesn't fit me.*
> *I don't like the colour.*
>
> **Making decisions**
> *Can I try it in a different size?*
> *I'll have this/these. I'll take it/them.*   *I'll leave it.*

1 Complete the sentences with a reflexive pronoun.

1 Was that T-shirt a gift? No, I bought it for _____.
2 Tim and I enjoyed _____ at the disco last night.
3 She doesn't live by _____ . She lives with friends.
4 Can I have some coffee? Yes, would you like to serve _____?
5 They taught _____ to speak Russian. They didn't have lessons.
6 He's unhealthy and smokes too much. He doesn't look after _____.

2 Work in pairs and act out the conversation in *Speaking and listening* activity 3.

VOCABULARY AND LISTENING

1 Look at the words in the box. Which items can you see in the photos?

> chocolates biscuits cakes flowers milk perfume jeans soap

бисќитс

2 Which words in 1 do the words below go with?

> box packet bottle bunch pair bar

гроз́ьб

a box of chocolates, ...

3 Work in pairs. Which items do you buy for yourself? Which items do you buy as gifts for other people?

I often buy chocolates for myself.

4 Listen to two conversations. Put the number of the conversation by the items in 1 which the people are buying.

обсужда́ть

5 Work in pairs and check your answers to 4. Who are they buying the items for?

Now listen again and check.

6 Work in pairs.

Student A: You're a shop assistant. You sell the items in the photos.
Student B: You're a customer. You want to buy something in the photos.

Act out a conversation. Use the conversations in 4 to help you.

SOUNDS

Listen to the conversations in *Vocabulary and listening* activity 4 again. In which conversation does the customer sound polite and friendly?

SPEAKING

1 Work in groups of three or four and discuss your answers to the questions.

1 Do you like shopping?
2 When you go shopping, do you usually go by yourself?
3 Do you know what you want to buy?
4 Do you usually buy things for yourself?
5 How often do you buy things for other people?
6 Who do you buy gifts for?
7 When do you give gifts?
8 What do you buy for gifts?

2 Find out what other students' answers are.

го́ловы - голла́у

3 Report back to your group.

Mario often buys things for himself.

27 | Whose bag is this?

Whose; possessive pronouns; describing objects

VOCABULARY AND SPEAKING

1 Complete the sentences below the pictures with
words from the box.

| short   long   square   round   small   large   rectangular |
| heavy   light |

rek'taNGyyələr.
прямоугольний

2 Say what the things in the pictures are made of.
Use the words in the box.

| glass   leather   metal   plastic   paper   wood |

конец
метəл

The ruler is made of plastic.

3 Work in pairs. Choose an object in the classroom. In
turn, ask and answer questions to try and guess the
object your partner is describing.

A *It's square.*
B *Is it a book?*
A *No. It's made of glass.*
B *Is it the window?*
A *Yes.*

4 Match the conversations and the pictures below.

A Excuse me, is this yours?   **A** Whose is this?
B No, it isn't mine. It's his.   **B** It's theirs.

a *It's long.*
b *It's short.*
c *It's square.*
d *It's_____ .*
e *It's_____ .*
f *It's_____ .*
g *It's_____ .*
h *It's_____ .*
i *It's_____ .*

GRAMMAR AND FUNCTIONS

> *Whose*
>
> **You use *whose* to ask who something belongs to.**
>
> ***Whose*** *bag is this?*
> ***Whose*** *shoes are these?*
>
> Possessive pronouns
>
> **You use possessive pronouns to say who something belongs to.**
>
> *mine yours his hers ours theirs*
> *Whose bag is this?* *It's* **mine.**
> *Whose shoes are these?* *They're* **his.**
>
> Describing objects
>
> **You don't usually put more than two or three adjectives together.**
>
> *What's it like?*
> *It's a small, plastic ruler.*

1 Look at these possessive adjectives.

my your his her its our their

Which letter do you add to the possessive adjective to make a possessive pronoun? Which possessive pronouns are the exception?

2 Choose the correct word in these sentences.

1 Whose is this? It's *my/mine*.
2 Where did I put *my/mine* bag?
3 These aren't *her/hers* shoes. They're *my/mine*.
4 *Whose/who's* got my pen?
5 Are these *their/theirs* books? No, they're *our/ours*.
6 *Whose/who's* coat is this?

3 Work in groups of four or five. Put two personal possessions on a desk or in a bag. Go round, in turn, holding up or taking out a possession, asking and saying who it belongs to.

A *Whose is this?*
B *It's mine. And whose is this?*
C *It's his.*

LISTENING AND SPEAKING

1 You're going to hear a conversation in a Lost Property office. Look at the form below. Match the items 1 – 9 and the questions a – i.

Lost property		
1 Name	Mrs Joan Fairfield	
2 Address	22, Burn Lane, Macclesfield	
3 Telephone	678 5463	
4 Lost article	bag	
5 Date of loss	20 July	
6 Time of loss	10am	
7 Place of loss	Chester market	
8 Description	small, square and it was made of black leather	
9 Contents	a purse, a calculator, an address book, a comb	

a What did you lose?
b Where did you lose it?
c What was in it?
d What date did you lose it?
e What's your name?
f What's your address?
g What's it like?
h What's your telephone number?
i What time did you lose it?

2 Look at the Lost Property form. Listen and underline any information which is different from what you hear.

3 Work in pairs. Correct any information which was different. Listen again and check.

4 Work in pairs. Act out the conversation you heard in 2. Use the form and the questions to help you.

5 Listen to three more conversations in a Lost Property office. Put the number of the conversation by the answer to the questions below. There is one extra answer for each question.

John Smith	Ian Joseph	Mary Walter	Ken Hamilton
13, Dock Lane, London	21, Tree Road, Leeds	33, James Street, Bath	45, Old Road, Oxford
56778	56983	75859	75889
bag	wallet	coat	box of cigars
20 March	31 May	12 June	17 October
11am	2pm	4pm	11pm
train	bus	supermarket	chemist
rectangular, nylon	black, plastic	red leather	wooden, square

6 Work in pairs. Act out the conversations in 5. Change your partner for each conversation.

Asking and saying how you feel; sympathising; *should, shouldn't*

симпфайдин
симпатизирующий

VOCABULARY AND LISTENING

1 Look at the words in the box. Find:

эраидоби
забопевание

– two types of medicine
– six complaints or illnesses
– five adjectives to describe how you feel
– seven parts of the body

потеря сознания
слабый
воспаленный

arm	aspirin	back	cold (noun)
cough	cough medicine	dizzy – *головокр*	
faint	finger	foot	hand
headache	ill	leg	sick
sore throat	stomach ache		
temperature	tired	toe ↓ O	

2 🔊 Listen to three conversations. Say what's wrong with each person.

The first person has got a headache.

3 Put the number of the person by what you think he/she should do.

- ☐ go to bed
- ☐ stay at home
- ☐ drink plenty of water
- ☐ stop smoking
- ☐ get some exercise
- ☐ go to the doctor
- ☐ keep warm
- ☐ eat nothing for 24 hours
- ☐ lie down

FUNCTIONS AND GRAMMAR

Asking and saying how you feel

What's the matter?   *I don't feel very well.*
Are you all right?   *I feel sick.*
           I've got a headache.
сй   *My back hurts.*

Sympathising

Oh dear!   *What a pity!*   *Oh, I am sorry!*

Should, shouldn't

Should and ***shouldn't*** are modal verbs.

You use ***should*** and ***shouldn't*** to give advice.

*You **should** go to bed.*   *You **shouldn't** go to work.*

(For more information about modal verbs, see Grammar review page 111.)

функции

1 Look at the functions and grammar box and answer the questions.

1 What follows *I feel* – an adjective or a noun?
2 What follows *I've got* – an adjective or a noun?
3 What comes before *hurts* a person or a part of the body?

2 Work in pairs and say what the people in the conversations in *Listening* activity 2 should or shouldn't do.

I think he should go to bed.
He shouldn't go to work.

3 Work in pairs.

Student A: Turn to Communication activity 16 on page 102.
Student B: Turn to Communication activity 21 on page 104.

READING AND WRITING

1 Read and answer the questions for your country.

1 When you're ill, do you go to a specialist who knows about your illness or your local doctor?
2 Are there both men and women doctors?
3 Where do you get medicine in your country?
4 Do you ever go to the doctor if you're well?
5 Do doctors visit you at home?
6 What do you do in an emergency?
7 Do friends and relatives visit you in hospital?
8 Do you pay for medical treatment? *лечении*
терапии
листовке

2 Read the advice leaflet *The nation's health* and find the answers to the questions for Britain.

The nation's health

In Britain, when you're ill, you go to a doctor near your home. Doctors are men and women, and you can say who you prefer. You usually only spend about ten minutes with the doctor. They can usually say what the matter is very quickly, and often give you a prescription for some medicine. You get this at the chemist's shop. If not, they may *npegraraiom* suggest you go to a specialist.

Most people only go to their doctor when they're ill. People with colds and coughs don't go to their doctor but to the chemist, to buy medicine. Doctors only come to your home when you're very ill. In an emergency you can call an ambulance on 999. The ambulance takes you to hospital for treatment. Friends and relatives visit you in hospital at certain hours of the day, but they don't stay there.

You don't pay for a visit to the doctor or to the hospital in Britain, but when you work you pay a government tax for your medical care. You also pay for prescriptions if you're over 18.

3 Look at the passage again and find a word which means the same as:

1 A special form for some medicine.
2 A doctor who knows a lot about an illness.
3 A place where there are many people who are ill.
4 A means of transport which takes you to hospital.

4 Work in pairs and discuss your answers to 3.

5 Work in pairs. Is medical care in your country different from Britain?

6 Prepare an advice leaflet about medical care for a foreign visitor to your country. Write answers to the questions in 1. Use the passage to help you.

When you're ill in Japan you should go to a specialist in your illness ...

29 *Country factfile*

Making comparisons (1): comparative and superlative forms of short adjectives

VOCABULARY AND SOUNDS

ə́ pə z it

1 Match the adjectives with their opposites in the box.

> big cold dry fast high hot low old
> slow small wet young

2 Which adjective does not go with the noun?

mountn —

1 country	hot cold dizzy
2 mountain	high big heavy
3 person	healthy young high
4 child	young low small

3 Which of the words for measurements in the box do you use to describe the following? *i e/G(k)*

*centigreig
no cropagyrur
uikaie —* area length temperature

> centigrade metre millimetre centimetre
> kilometre square kilometre

4 🔲 Listen to the words in the box in 3 and underline the stressed syllable.

Now say the words aloud.

сокращение

5 Match the words for measurements and their abbreviations. *абривидиис.*

m mm km cm sq km °C

READING AND LISTENING

*информация о центре
в стране срактов.*

1 Look at the country factfile and decide if these statements are true or false.

1 Thailand is smaller than Britain.
2 It's colder in Thailand than in Britain.
3 It's drier in Thailand than in Britain.
4 The population in Thailand is smaller than in Britain.
5 The Thai armed forces are smaller than the British armed forces.
6 Thai children are younger when they start school than British children.
7 Thai children are older when they leave school than British children.

Country factfile

*'ev/p/r/j
konuecbo
oca geol*

		Kingdom of Thailand	United Kingdom of Great Britain and Northern Ireland	Sweden
1	Land area	513,115 sq km	242,429 sq km	
2	Average temperature	January 25°C July 28°C	January 4.5°C July 18°C	
3	Rainfall	1400 mm	600 mm	
4	Population	(1993) 58,722,437	(1993) 57,970,200	
5	Armed Forces	295,000 troops	274,800 troops	
6	Education	Free and compulsory for all	Free and compulsory for all	

*kəm'pəl zərē
обязательное*

2 You're going to hear Karl answering questions about Sweden. Listen and put the letter corresponding to the correct answer in the chart.

1 a 449,964 sq km   b 500,200 sq km

2 a January -3°C, July 18°C   b January 20°C, July 20°C

3 a 535 mm   b 450 mm

4 a 8,730,289   b 31,645,896

5 a 200,500 troops   b 64,800 troops

6 a All children between 7 and 16
   b All children between 7 and 17

3 Work in pairs. Can you remember what else Karl mentions?

Listen again and check.

GRAMMAR AND FUNCTIONS

kəm'para səns

> Making comparisons (1): comparative and superlative forms of short adjectives
>
> **You form the comparative of most short adjectives with -er, and the superlative with -est.**
>
> **adjective:** *old* *large* *big* *dirty*
> **comparative:** *older* *larger* *bigger* *dirtier*
> **superlative:** *oldest* *largest* *biggest* *dirtiest*
>
> **There are some irregular comparative and superlative forms.**
>
> *good* *better* *best*
> *bad* *worse* *worst*
>
> **You use a comparative adjective + *than* when you compare two things which are different**
>
> *Thailand is **bigger than** Britain.*

1 How do you form comparative and superlative adjectives ending in -e, -y, vowel + -d, -g, -m, -n or -t?

2 Write the comparative and superlative forms of these adjectives.

large fine close wide dirty dry healthy heavy
noisy big hot wet

3 Correct any statements in *Reading and listening* activity 1 which are false.

Thailand is bigger than Britain.

4 Complete these sentences with the adjective in brackets.

1 Britain is _____ than Sweden. (small)
2 Thailand is _____ than Sweden. (hot)
3 Britain is _____ than Thailand. (dry)
4 Thailand has the _____ armed forces of the three countries. (big)
5 The children in Sweden are _____ when they start school than the children in Thailand. (old)
6 Sweden is the _____ of the three countries. (cold)

WRITING

Write a factfile for your country. Use the chart in *Reading and listening* activity 1
OR
Write a paragraph comparing your country with Thailand, Britain or Sweden. It doesn't matter if you don't know exact figures. *фигорс*

I think my country is larger than Thailand, but the population is smaller.

30 Olympic spirit

**Making comparisons (2): comparative
and superlative forms of longer adjectives**

VOCABULARY AND LISTENING

1 Work in pairs. Which of these sports
can you see in the photos?

> football   motor racing   swimming
> tennis   golf   horseriding   climbing
> windsurfing   basketball   skiing
> hang gliding   cycling

мотогонки
езда на лошади
skeing - катание на лыжах
дельтапланеризм   на велике

Turn to Communication activity 28 on
page 105 and check you know what
the other sports are.

2 Put the words for sports in two
columns: *team sports* and *individual
sports.*

калайс

team sports: football
individual sports: swimming

Now work in pairs and check your
answers.

3 Match the adjectives in the box with
the sports in the vocabulary box in 1.

> popular   expensive   tiring   dangerous
> fashionable   difficult   exciting

модный

football: popular

Now work in pairs and find out if
your partner agrees with you.

4 Look at the statements about sport in the
chart. Tick (✓) the statements you agree with.

5 🔲   Listen to Katy and Andrew talking about their opinions about
sport. Tick (✓) the statements they agree with.

6 Work in pairs and check your answers to 5.
🔲   Listen again and check.

GRAMMAR AND FUNCTIONS

> **Making comparisons (2): comparative and superlative forms of
> longer adjectives**
>
> **You form the comparative of many long adjectives with
> *more* + adjective, and the superlative with *most* + adjective.**
>
> | adjective: | *expensive* | *tiring* |
> |---|---|---|
> | **comparative:** | ***more expensive*** | ***more tiring*** |
> | **superlative:** | ***most expensive*** | ***most tiring*** |
>
> *Motor racing is the **most exciting** sport in the world.*
> *Climbing is **more difficult** than skiing.*

	You	Katy	Andrew
The most popular sport is football.			
Horseriding is more expensive than cycling.			
Tennis is the most tiring sport.			
Hang gliding is more dangerous than windsurfing.			
Climbing is more difficult than skiing.			

1 Complete the sentences using the comparative or superlative form of the adjective.

1 Horseriding is very expensive.
Yes, it's _____ sport I can think of.

2 Motor racing is very dangerous.
Yes, it's _____ than skiing.

3 Football is very popular.
Yes, it's _____ than tennis.

4 Windsurfing is very difficult.
Yes, it's one of _____ sports I can think of.

5 Swimming is very tiring.
Yes, it's the _____ sport in the world.

2 Choose the correct sentence. Can you explain why?

1 a Football is one of the most popular games in the world.
b Football is one of the popularest games in the world.

2 a Skiing is the most difficult sport to do well.
b Skiing is the more difficult sport to do well.

3 a Britain is worse at tennis than many countries.
b Britain is worst at tennis than many countries.

3 Work in pairs. Find out how your partner completed the chart in *Vocabulary and listening* activity 4.

I think hang gliding is the most dangerous sport.
Do you? I think motor racing is more dangerous than hang gliding.

4 Write sentences using the comparative or superlative form of these adjectives.

interesting   lively   boring
intelligent   successful   enjoyable

The most interesting game in the world is chess.

SPEAKING AND READING

1 With the rest of the class, make a list of Olympic sports.

swimming, athletics, ...

2 Put these paragraphs from a story about the Olympic Games in the right order.

> **A** He then caught the train back to Stockholm, made a reservation into a hotel, got a boat to Japan, got married, had six children and ten grandchildren.
>
> **B** In 1966 Shizo Kanakuri finished the Olympic marathon in record time. To run the 42 kilometres, he took 54 years, 8 months, 6 days, 8 hours and 32 minutes.
>
> **C** Then he went back to the place where he stopped for a drink in 1912 and finished the marathon for Japan.
>
> **D** He started in 1912 in Stockholm, and after a few miles he saw some people having a drink. He was thirsty too, so he joined them.
>
> Adapted from *The Return of Heroic Failures*, by Stephen Pile

3 Choose the best title for the story.

1 The most expensive game   2 The slowest run
3 The worst match

4 Work in groups of two or three. Find out if people in your class enjoy the Olympic Games. If they enjoy them, what do they like and why? If they don't enjoy them, what do they dislike and why?

Do you like the Olympic Games?
What is the most enjoyable game?
What is the most boring game?

Progress check 26–30

VOCABULARY

1 When you write down an adjective make a note of its opposite meaning.

hot – cold   *light – dark*

Match the adjectives and their opposites.
(There may be more than one possibility.)

heavy   large   light   long   round   short   small   square

You may like to look through Lessons 1 to 30 and see if there are other adjectives and their opposites.

2 Work in groups of three or four and play *Word Zigzag* with words from Lessons 26 to 30.

How to play Word Zigzag

1 On a large sheet of paper, Student A writes a word from Lessons 26 to 30 horizontally.

2 Student B thinks of a word which includes one letter from Student A's word and writes it vertically.

3 Student C thinks of another word which includes a letter from Student B's word and writes it horizontally.

4 The game continues until no one can think of a suitable word. The last student to write a word is the winner.

```
                        straight
m o u s t a c h e
                        o
                        r
                        t
```

GRAMMAR

1 Choose the correct word in these sentences.

1 Is this *my/mine* pen? No, it's *my/mine*.
2 Whose are these? They're *her/hers*.
3 Where did you leave *your/yours* coat?
4 This isn't *their/theirs*. It's *your/yours*.
5 Have you got *my/mine* ticket?
6 These are *my/mine* gloves, not *your/yours*.

2 Write the comparative and superlative forms of the following adjectives.

large   good   popular   big   ridiculous   healthy   safe   wet   expensive   tiring   heavy   high   difficult

large   larger   largest

3 Complete these sentences with the comparative form of the adjective in brackets.

1 France is _____ than the USA. (small)
2 Parachuting is _____ than skiing. (dangerous)
3 Hang gliding is _____ than windsurfing. (expensive)
4 Winter in Norway is _____ than winter in Brazil. (cold)
5 Football is _____ than boxing. (popular)
6 Cycling is _____ than hang gliding. (exciting)

4 Complete the sentences using the comparative or superlative form of the adjective.

1 Russia is a very large country. Yes, it's the _____ country in the world.
2 Motor racing is very expensive. Yes, it's the _____ sport I can think of.
3 Switzerland is quite a small country. Yes, it's _____ than Britain.
4 Swimming is a tiring sport. Yes, it's _____ than cricket.
5 Football is a very popular sport. Yes, it's the _____ sport in the world.

5 Write two words to complete the following sentences.

1 I feel _____.
2 My _____ hurt/hurts.
3 I've got _____.

6 Reply to these people and give advice. Use *should/shouldn't*.

1 I'm tired.
2 I've got toothache.
3 My back hurts.
4 I feel sick.
5 I've got a cold.
6 I've got a cough.

SOUNDS

1 Group the words which rhyme.

could scarf half said head wood red laugh

Listen and check. As you listen, say the words aloud.

2 Say these words aloud. Is the underlined sound /eɪ/ or /aɪ/?

f<u>a</u>ce f<u>i</u>ne s<u>ig</u>n M<u>ay</u> Sp<u>ai</u>n n<u>ig</u>ht l<u>ie</u> m<u>ai</u>d tr<u>ay</u>

Listen and check. As you listen, say the words aloud.

3 Match the words and the stress patterns.

☐ ☐ ☐   ☐ ☐ ☐ ☐

reception caravan camera rectangular receiver
umbrella temperature

Listen and check. As you listen, say the words aloud.

4 Underline the words you think the speakers will stress.

CUSTOMER	And I must have lost it then.
OFFICIAL	Just say your name again, madam.
CUSTOMER	Mary Walter.
OFFICIAL	And your address and phone number?
CUSTOMER	21, Tree Road, Leeds, 75889.
OFFICIAL	And it was a black plastic bag, you say?
CUSTOMER	Yes.
OFFICIAL	And you last saw it on 20th March at two in the afternoon.
CUSTOMER	Yes, in the supermarket.
OFFICIAL	And what was in it?
CUSTOMER	All my shopping and my purse.

5 Listen and check. Which speaker sounds interested?

Now work in pairs and act out the conversation. Try to sound interested.

SPEAKING

1 Work in groups of three or four. Look at these sentences and decide what the situation is.

1 I've got a sore throat.
2 Have you got it in red?
3 When did you lose it?
4 My back hurts.

2 Match the sentences in 1 with the replies below.

a This morning.
b Not in your size, we haven't.
c Yes, it does look a bit red.
d What have you done?

3 Choose one or two conversations and write a few sentences to continue them. When you're ready, act them out to the rest of the class.

31 | *When in Rome, do as the Romans do*

Needn't, can, must, mustn't

VOCABULARY AND READING

1 Work in pairs. Use the words in the box to say what's happening in the photos. *указывать на*

> shake hands   cover   point at
> kiss   take off

2 Read *When in Rome* and match the rules and advice with the photos.

When in Rome

- [ ] In parts of Africa you must ask if you want to take a photograph of someone.
- [ ] In Japan you must take off your shoes when you go into someone's house.
- [ ] In Saudi Arabia women must cover their heads in public.
- [ ] In Britain you mustn't point at people.
- [ ] In Japan you mustn't look people in the eye.
- [ ] In China you mustn't kiss in public.
- [ ] In Taiwan you must give a gift with both hands.
- [ ] In France you must shake hands when you meet someone.

LISTENING

1 🔲 Listen to James, who's Australian, talking about some of the advice and rules in *When in Rome*. Tick (✓) the statements he talks about.

2 Look at these sentences.

Women needn't cover their heads in Australia.

In Australia you can look people in the eye.

Complete these sentences with *needn't* and *can* so that they are true for Australia.

1 You _____ ask if you want to take a photograph of someone.

2 You _____ take your shoes off when you go into someone's house.

3 You _____ kiss in public in Australia.

4 You _____ shake hands with everyone when you meet them in Australia. You _____ shake hands when you meet someone for the first time.

3 🔲 Now listen again and check.

GRAMMAR

Needn't and can

You use *needn't* if it isn't necessary to do something.

*Women **needn't** cover their heads in Australia.*

You use *can* if you're allowed to do something.

*In Australia you **can** look people in the eye.*

***Can* is also a modal verb. (For more information about modal verbs see Grammar review page 111.)**

Must, mustn't

***Must* and *mustn't* are modal verbs.**

You use *must* to talk about something you're strongly advised to do or are obliged to do, such as rules. *обладжед*

*You **must** ask if you want to take a photograph of someone.* (strong advice)

*Women **must** cover their heads in public.* (rule)

You use *mustn't* to talk about something you're strongly advised not to do or are not allowed to do. *— nozbouem*

*You **mustn't** point at people.* (strong advice)

*You **mustn't** kiss in public.* (rule)

1 Complete these sentences with *must* or *mustn't*.

1 Children _____ play near the road.

2 You _____ be quiet in a library.

3 You _____ keep your wallet in a safe place.

4 Men _____ take off their hats in a church.

5 You _____ give a gift with one hand in Taiwan.

6 You _____ wear shoes in a Japanese home.

2 Write some strong advice for people learning a foreign language.

You must come to classes every week.
You mustn't miss any lessons.

3 Write some rules for your school or the place where you work.

You mustn't smoke during lessons.

SOUNDS

1 🔲 Listen to the way you pronounce *must* /mʌst/ and *mustn't* /mʌsnt/. Look at the sentences in *Grammar* activity 1.

2 Say the sentences aloud. Make sure you pronounce *must* and *mustn't* correctly.

SPEAKING AND WRITING

1 Work in pairs. Think of strong advice and rules you can give to visitors to your country about the following:

– giving presents   – what to wear
– home visits   – eating habits
– table manners

When you receive a gift in Spain, you must open it immediately.

2 Work with another pair. Do you have similar rules and advice? Write a list of advice and rules for visitors to your country.

32 | *Have you ever been to London?*

Present perfect (1): talking about experiences

READING AND VOCABULARY

знаменитая достопримечательность

1 Work in pairs. Make a list of famous sights to see in your town or capital city.

2 Look at these famous sights of London. Do you know what they are?

3 Read this postcard from London. Which photo in 2 is on the back of it?

4 Here is a list of some of the things you can do in London. Tick (✓) the places Guy and Emma have been to or the things they've done.

- ☐ watch the Changing of the Guard
- ☐ visit Westminster Abbey
- ☐ visit St Paul's Cathedral
- ☐ listen to a concert in St James' Park
- ☐ climb Tower Bridge
- ☐ go to Hampstead

Dear David and Anna,

Hi! How are you? We're having a wonderful time in London. We're staying in a hotel in the centre of London. We've only been here four days but we've done so much already. We've watched the Changing of the Guard at Buckingham Palace and we've listened to a concert in St James' Park. We've visited St Paul's Cathedral, but not Westminster Abbey. We've climbed Tower Bridge (you can see it on this postcard) and we've been to Greenwich by boat, but we haven't been to Hampstead yet. We're going there tomorrow.

See you soon!

Love Guy and Emma

David and Anna Sayle
Apartment 214
51st West City Street
New York 10021
USA

5 Complete the sentences below with the words in the box.

| climb cathedral bridge boat park view |

1 A _____ is the most important church in a city.
2 You cross a river by going over a _____ or by taking a _____.
3 When you _____ something, you go up it.
4 A _____ is a public place in a town, with trees and grass.
5 The _____ from a building is what you can see from it.

SOUNDS

1 🔊 Listen and repeat.

ever   Have you ever   Have you ever been
Have you ever been to London?
ever   Have you ever   Have you ever stayed
Have you ever stayed in a hotel?

2 🔊 Read and listen to this conversation. Underline the stressed words.

A Have you ever been to London?
B No, I haven't. I've never been there.
A Have you ever stayed in a hotel?
B Yes, I have.
A When was that?
B When I was in Spain last year.

3 Work in pairs and practise the conversation in 2.

GRAMMAR

Present perfect (1): talking about experiences
You use the present perfect to talk about an experience, often with *ever* and *never*.
***Have you ever stayed** in a hotel?*
(= Do you have the experience of staying in a hotel?)
Yes, I have. (= Yes, at some time in my life, but it's not important when.)
*No, I haven't. **I've never stayed** in a hotel.*
You form the present perfect with *has/have* + past participle. You usually use the contracted form, *'ve* or *'s*.
Many past participles are irregular.
Have you ever been to London?   We've done so much.
For a list of irregular participles, see page 114.
Remember that you use the past simple to talk about a definite time in the past.
When did you stay in a hotel?   When I was in Spain last year.

1 Work in pairs. Ask and answer questions about what Guy and Emma have done on their holiday.

1 watch the Changing of the Guard
2 visit St Paul's Cathedral
3 climb Tower Bridge
4 visit Westminster Abbey
5 listen to a concert in St James' Park
6 go to Hampstead

1 Have they watched the Changing of the Guard? Yes, they have.

2 Here are the regular past participles of some verbs. Write the infinitive.

lived   worked   stayed   watched   visited   listened

3 Match the infinitives with their irregular past participles.

eat drink drive read see fly take buy win make write send

driven sent read seen flown bought won made eaten drunk taken written

SPEAKING AND WRITING

1 Think of things to do and places to see in the town where you are now. Go round the class and ask and say what people have done in your town.

Have you seen the cathedral?   Yes, I have.
Have you taken the boat along the river?
No, I haven't.

2 Imagine you're a visitor to your town. Write a postcard to a friend saying what you've seen and where you've been. Use the postcard in *Reading* activity 3 to help you.

Dear Enrique,
Hi, how are you? I'm in Seville at the moment. I've seen the Alcazar...

Present perfect (2): talking about recent events; *just* and *yet*

недавний

LISTENING AND VOCABULARY

1 Match these sentences with the pictures of Barry.

He's hurt his back.   He's lost his wallet.   He's failed his exam.

2 Look at this conversation. Can you guess what Barry says?

ALAN   Hi! How's your day been?
BARRY   (1) _____
ALAN   I'm sorry to hear that. What's happened?
BARRY   (2) _____
ALAN   Your back! How did you hurt it?
BARRY   (3) _____
ALAN   A box of books! I'm not surprised you hurt yourself trying to lift a box of books. Have you been to the doctor yet?
BARRY   (4) _____
ALAN   Well, I think you should go immediately. And what else has happened?
BARRY   (5) _____
ALAN   Your wallet? Where did you lose it?
BARRY   (6) _____
ALAN   Have you been back to the bus stop yet?
BARRY   (7) _____
ALAN   And have you heard your exam result?
BARRY   (8) _____
ALAN   Have you passed?
BARRY   (9) _____
ALAN   Oh dear, it's been one of those days for you, hasn't it?

3 📼 Listen and check your answers to 2.

4 Work in pairs and act out the conversation.

5 Match the verbs and the nouns in the box. There may be several possibilities.

| break cut plate goal drop bag miss wallet lose pass fail |
| arm pay score steal bill exam finger train crash car catch |

break – arm, finger...

6 🔊 Listen to four more conversations. Put the number of the conversation by the verbs in 5 which you hear.

GRAMMAR

> Present perfect (2): talking about recent events; *just* and *yet*.
>
> **You use the present perfect to talk about recent events, such as a past action which has a result in the present. You often use to it describe a change.**
> *He's hurt his back.   He's lost his wallet.*
> **You use *just* if the action is very recent.**
> *He's just lost his wallet.*
> **You use *yet* in questions and negatives to talk about an action which is expected.**
> *Have you been to the doctor yet?*
> *I haven't gone back to the bus stop yet.*
> **Remember that you use the past simple to say when the action happened.**
> *When did you lose your wallet? I lost it this morning.*

1 Write sentences saying what has happened in the conversations in *Listening and vocabulary* activity 6.

In the first conversation, someone has stolen her bag.

2 Look at the example sentences in the grammar box. Where do you put *just* and *yet* in a sentence?

3 Choose the correct sentence. Can you explain why?

1 a *I've seen* her yesterday.
  b *I saw* her yesterday. √
2 a *Have you been* shopping yet? √
  b *Did you go* shopping yet?
3 a *She's lost* her wallet last week.
  b *She lost* her wallet last week. √
4 a *I've never been* to England. √
  b *I never went* to England.
5 a *We've just decided* where to go on holiday. √
  b *We just decided* where to go on holiday.
6 a *Have you met* anyone famous when you were in Hollywood?
  b *Did you meet* anyone famous when you were in Hollywood? √

SPEAKING

1 Make a list of things you've done and things you haven't done yet this week.

call my mother, pay the bills, do some shopping, write to the bank manager, . . .

2 Work in pairs. Show each other the lists you made in 1. Ask and say what you've done and haven't done yet.

Have you called your mother yet?
Yes, I have.
Have you paid the bills yet?
No, I haven't.

3 Work in pairs. Act out the following situations.

Student A: Look at the statements below and tell Student B what's happened using the present perfect. Answer his/her questions.

– you break your arm
– you find some money
– you lose your friend's pen
– you fail your exam
– you drop an expensive plate

I've broken my arm!

When you've finished, react to Student B's situations. Ask questions using the past simple.

Student B: React to Student A's situations. Ask questions using the past simple.

STUDENT A   *I've broken my arm!*
STUDENT B   *How did you do that?*
                   Where did it happen?

When you've finished, tell Student A what's happened to you. Remember to use the present perfect tense.

– you win a million pounds
– you cut your finger
– you miss the last bus home
– you burn your dinner
– you crash the car

SPEAKING

1 Work in pairs. What is your idea of a perfect day out? Here are some suggestions:

– a shopping trip
– a visit to the beach
– a picnic in the country
– a visit to a historical building
– a visit to some friends
– a walk in the mountains

2 Look at the photo. Which of the situations in 1 does it show?

READING AND VOCABULARY

1 Work in pairs. Make a list of things to do or take on a perfect picnic.

2 Read *The perfect picnic* and decide which paragraphs the pictures illustrate.

3 Match these words with the pictures.

bottle opener   barbecue matches   knife   fork   ice cup   carton   rubbish blanket *одеяло*

4 Complete these sentences with words from the passage.

1 I was very _____ so I had something to eat.
2 The weather was _____ so we had dinner outside.
3 When you go away for a night or two, don't forget to _____ your toothbrush.
4 Check the weather _____ to find out if it's going to rain.
5 I was cold in bed so I asked for another _____.

5 Work in pairs. Do you agree with the advice in the passage? Do you have picnics like this in your country?

The perfect picnic

Everyone says that food and drink taste better when you have a picnic. But what do you do to have a perfect picnic? Here's some advice.

1 Choose where you want to go very carefully. In the country? In the city? The picnic site should be attractive and interesting, to be sure there's plenty to do when you finish your picnic.

2 Check the weather forecast the day before you go. The perfect picnic needs perfect weather.

3 Don't take too much to carry. For the perfect picnic you leave home with food and drink and you return only with rubbish.

4 Choose small items of food, such as eggs or sandwiches, to avoid taking knives and forks. To make it the perfect picnic, take food which you don't usually eat.

5 Take small cartons of juice or plastic bottles of water. They're more expensive, but they aren't as heavy as glass bottles, cups and glasses.

6 Pack a blanket to sit on or, if it's cold, to keep you warm.

7 Put fresh food in a bag with ice to keep it cool.

8 Put the whole picnic in a number of small bags, to allow everyone to carry something.

9 Prepare everything before you go OR make sure you've got everything you need to finish preparing the picnic, such as a knife, a bottle opener, barbecue, matches.

10 Check there is a short walk to the picnic site to make people hungry.

GRAMMAR

> **Imperatives**
>
> **You use an imperative (infinitive without *to*) to give instructions and advice.**
> ***Check** the weather forecast.*
>
> **You use *don't* + imperative for a negative instruction.**
> ***Don't take** too much to carry.*
>
> Infinitive of purpose
>
> **You use *to* + infinitive:**
>
> **– to say why people do things.**
> *Try to have your picnic on a weekday, **to avoid** the weekend traffic.*
>
> **– to say what you use something to do.**
> *You use a bottle opener **to open** bottles.*

1 Answer the questions. Use *to* + infinitive.

1 Why should the picnic site be attractive and interesting?
2 Why should you choose small items of food?
3 Why should you pack a blanket?
4 Why should you put fresh food in a bag with ice?
5 Why should you put the whole picnic in a number of small bags?
6 Why should you check there's a short walk to the picnic site?

1 To be sure there's plenty to do when you finish your picnic.

2 Match words in the vocabulary box in *Reading and vocabulary* activity 3 with what you use them to do.

1 to put food in your mouth   4 to keep something cold
2 to cook food outdoors   5 to light a barbecue
3 to cut food   6 to sit on or to keep you warm

3 Work in pairs and check your answers to 2.

You use a fork to put food in your mouth.

4 Think of reasons why you do the following. Use *to* + infinitive.

1 go shopping   4 go to work/school
2 take a holiday   5 go to the cinema
3 go to the airport 6 use a knife

1 You go shopping to buy food or clothes.

WRITING

1 Work in groups of two or three. You're going to prepare some advice for planning the perfect day out. Make sure you all choose the same situation from *Speaking* activity 1.

First, work alone. Make notes on your advice.

a shopping trip – make a list

2 Work with the rest of the group. Make a list of all your advice, and explain why. Use an imperative and *to* + infinitive.

Make a list of things to buy to be sure you don't forget anything.

3 Show your instructions to another group. Do they agree with your advice? Is there anything which surprises them?

35 | *She sings well*

Adverbs

Music	She sings well and she plays the piano and guitar beautifully.
Art	She draws very carefully. Her work is excellent.
General	She is a very artistic young woman.

Antonia

VOCABULARY AND SOUNDS

1 Match the adverbs and their opposites in the box below.

> badly carefully carelessly quickly
> loudly politely quietly rudely
> slowly well

Sport	He can run quickly, and plays tennis well. He's good at most sports.
English	He writes very slowly and his spelling is very bad
General	He finds many subjects very difficult. Must try harder.

Guillaume

2 📼 Listen to four conversations. Which adverbs would you choose to describe how the speakers are speaking?

READING AND SPEAKING

Maths	She can add and subtract numbers quickly in her head.
Science	She is good at biology. She passed the exam easily.
General	Beate shows great ability in her work.

Beate

1 Work in pairs. Look at the photos. What do you think the people do?

2 Work in pairs. Who do you think was good at school?

French	She speaks french almost fluently. Well done!
English	She writes English compositions confidently. She works very hard.
General	Her manners are excellent. She talks to people quietly and politely.

Kate

3 Work in pairs. Read the extracts from school reports and match the adult in the photo with the child in the report.

4 Say what the people were good at when they were at school.

Antonia was good at music.

5 Work in pairs and say what you were good at when you were at your first school.

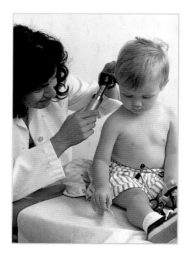

GRAMMAR

> **Adverbs**
> **You use an adverb to describe a verb.**
> *She speaks **slowly**.*
> **You usually form an adverb by adding -ly to the adjective.**
> *quiet – quietly*
> *loud – loudly*
> **If the adjective ends in -y, you drop the -y and add -ily.**
> *easy – easily*
> **Some adverbs have the same form as the adjective they come from.**
> *late   early   hard*
> **The adverb from the adjective good is well.**
> *She's a **good** singer. She sings **well**.*

1 Write the adjectives which the adverbs in the vocabulary box come from.

badly – bad

2 Write the adverbs which come from these adjectives.

angry   happy   gentle   quick
immediate   successful   comfortable
sudden   funny   frequent

3 Complete the sentences with the adjective in brackets or the adverb which comes from it.

1 He spoke _____ so everyone could hear him _____.
(clear, good)

2 They were late so they had a _____ game of tennis and then left. (quick)

3 She had a very _____ lesson with her pupils. (successful)

4 He listened to his teacher very _____. (careful)

5 Could you speak more _____, please. Your accent is _____ to understand. (slow, hard)

6 He passed the spoken exam very _____. (easy)

4 Work in pairs. Say what people in your class can do and how well they do it. Use an adverb.

Guido can run quickly.

LISTENING AND SPEAKING

1 Think about your answers to these questions about school.

	You	Gavin	Jenny
Do/did you always work very hard?			
Do/did you always listen carefully to your teachers?			
Do/did you always behave very well?			
Do/did you pass your exams easily?			
Do/did you always write your homework slowly and carefully?			
Do you think schooldays are/were the best days of your life?			

2 🔲 Listen to Gavin and Jenny, who are English, answering the questions. Put a tick (✓) by the ones they say *yes* to.

3 Work in pairs and check your answers to 2. Can you remember what Gavin and Jenny said in detail?

🔲 Now listen again and check.

4 Work in pairs. Ask your partner the questions in the chart.

5 Work in groups of three or four.

Student A, B, C: Student D has chosen an adverb from the vocabulary box. Ask him/her to:

– say something   – sing
– perform an action   – read something from *Reward* Elementary

He/she will perform the action in the manner of the adverb. Try to guess the adverb.

Student D: Choose an adverb from the vocabulary box. Don't tell the others what it is. The students in your group will ask you to do something in the manner of the adverb you have chosen. They must guess the adverb you have chosen.

Change round when you're ready.

Progress check 31–35

VOCABULARY

словосочетание
2 эээ

1 A collocation is two or more words which often go together.

fast car   high mountain   busy street   have dinner

Here are some adjectives from Lessons 31 to 35.

cold   difficult   low   old   expensive

Think of nouns which often go with the adjectives. You can use your dictionary.

cold day

2 There may be many places outside your classroom where you can see and listen to English, and build your vocabulary. In which of the following places can you see or listen to English words?

– food labels   *пищевые этикетки*
– the airport
– notices and signs
– instructions (for electrical goods) *электротовара*
– the station
– the radio
– travel documents (tickets etc)
– the television
– newspapers

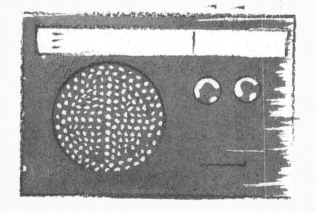

GRAMMAR

1 Complete these sentences from Lesson 31 with *must* or *mustn't*.

1 In parts of Africa you _____ ask if you want to take a photograph of someone.
2 In Saudi Arabia women _____ cover their heads in public.
3 In China you _____ kiss in public.
4 In Japan you _____ look people in the eye.
5 In Taiwan you _____ give a gift with both hands.
6 In Britain you _____ point at people.

2 Write sentences saying why you:

1 use a blanket   4 use a match
2 go to the swimming pool 5 use a bottle opener
3 go to the supermarket   6 use the telephone

1 You use a blanket to keep warm.

3 Write the adverbs which come from these adjectives.

easy   good   careful   fast   hard   polite   quiet   rude

4 Complete the sentences with the adjective in brackets or the adverb which comes from it.

1 She speaks English very _____. (good)
2 He drives extremely _____. (fast)
3 He's extremely _____ to people. (polite)
4 Don't make so much noise. Please be _____. (quiet)
5 Hungarian is a _____ language to learn. (hard)
6 He passed his First Certificate exam _____.(easy)

SOUNDS

1 Group the words with the same vowel sound.

beer stair hair near hear chair air we're year

 Listen and check. As you listen, say the words aloud.

2 Say these words aloud. Is the underlined sound /əʊ/ or /ɔː/?

g<u>o</u> s<u>o</u> J<u>o</u> l<u>aw</u> t<u>o</u>re w<u>ar</u> fl<u>oo</u>r l<u>ow</u> s<u>o</u>re t<u>oe</u> sp<u>or</u>t

 Listen and check. As you listen, say these words aloud.

3 Listen to how you can change the stressed word in a question and get a different answer.

1 a Can you **speak** Spanish?
   No, but I can **write** it.
  b Can you speak **Spanish**?
   No, but I can speak **Italian**.

2 a Did you stay with friends in **Paris**?
   No, I stayed with friends in **Rome**.
  b Did you stay with **friends** in Paris?
   No, I stayed in a **hotel**.

3 a Have you got this **dress** in another colour?
   No, only the **jeans**.
  b Have you got this dress in another **colour**?
   No, we've only got it in **red**.

READING AND WRITING

1 Look at the pictures below. Can you guess what happens in the story. Now read the story and see if you guessed correctly.

2 Read the story again and cross out any words which aren't 'necessary'. You cannot cross out two or more words together.

A young man went into a local bank, went up to the woman cashier and gave her a note and a plastic bag. The note said, 'Put all your money into this bag, please.' The middle-aged cashier was very frightened so she gave him all the money. He put it in his bag and ran out of the front door. When he got back home the city police were there. His note was on an old, white envelope and on the envelope was his home address.

3 Work in pairs and check your answers to 2.

4 Work in pairs and choose a short passage from *Reward* Elementary.

Working alone, count the number of 'unnecessary' words in the passage.

Tell each other how many words you've found. Who has found the most?

36 *I'll go by train*

Future simple (1): *(will)* for decisions

VOCABULARY AND SOUNDS

1 Look at the words in the box. Put them under two headings: *train* and *plane*. Use a dictionary, if necessary.

departure lounge   passport control   baggage reclaim
check-in   arrival hall   platform   boarding pass
business class   first class   cheap day return   return
single   departure gate   ticket office   tourist class

2 Which is the odd word out?

1 departure lounge   business class   check-in
2 platform   ticket office   departure gate
3 tourist class   first class   arrival hall
4 ticket office   passport control   boarding pass

3 🔲 Listen to these two-word nouns. Underline the stressed word.

departure lounge   passport control   business class
ticket office   arrival hall

Now say the words aloud.

LISTENING

1 Look at this conversation. Where does it take place?

A Can I help you?
B Yes, I'd like a ticket to London.
A When do you want to travel? It's cheaper after 9.15.
B I'll travel after 9.15.
A Single or return?
B I'll have a cheap day return ticket, please.
A That'll be thirteen pounds exactly. How would you like to pay?
B Do you accept credit cards?
A I'm afraid not.
B Well, I'll pay cash, then. Will there be refreshments on the train?
A No, I'm afraid there won't.
B Can I have a ticket for the car park as well.
A That'll be fifteen thirty in all.
B Thank you.

2 🔲 Listen and underline anything which is different from what you hear.

GRAMMAR

> **Future simple (1):** *(will)* for decisions
> **You form the future simple with** *will* **or**
> *won't* **+ infinitive.**
>
> | I | | |
> |---|---|---|
> | you | | |
> | he/she/it | 'll (will) | |
> | we | won't (will not) | go by train. |
> | they | | |
>
> **You use** *will* **when you make a decision at the time of**
> **speaking.**
> *I'll have a return ticket.* *I'll pay cash.*

1 Work in pairs.

Student A: You're going to act out the conversation in
Listening activity 1. You're in the ticket office. Change
some of the details in the conversation.

Student B: You're going to act out the conversation in
Listening activity 1. You're the passenger. Change
some of the details about the ticket you want to buy.

2 Act out the conversation when you're ready.

READING AND SPEAKING

1 Read the travel brochure and follow the route on
the map.

2 Imagine you want to go on the tour described in the
travel brochure. Look at the brochure and answer the
questions.

  1 How long will you spend on the train?
  2 How long will you spend in Cape Town?
  3 Will you have to pay extra for meals?
  4 How will you get from Livingstone to Windhoek?
  5 Will you be able to have a single room?
  6 Will you be able to go to a game reserve?
  7 Will there be anything extra to pay?
  8 Will you be able to stay an extra day?

3 Work in groups of three.

Student A: Turn to Communication activity 19 on
page 104.

Student B: Turn to Communication activity 9 on
page 100.

Student C: Turn to Communication activity 11 on
page 101.

TRAVEL IN STYLE –
FROM THE CAPE TO VICTORIA FALLS

Visiting Cape Town, Kimberley, Pretoria and the Victoria Falls.
Including a trip on the Pride of Africa – probably the finest train in
the world!

FOR ONLY £2595 per person

ITINERARY

Day 1 Leave London and fly overnight to Cape Town.
Day 2 – 4 Sightseeing in Cape Town.
Day 5 – 7 Join the train and travel overnight to Kimberley, and on
to Johannesburg and Pretoria.
Day 8 Leave Pretoria by train.
Day 9 Travel all day across Zimbabwe, and on to the Victoria Falls.
Day 10 – 12 Leave the train. Sightseeing around the Victoria Falls
and a visit to Chobe game reserve.
Day 13 Take the plane from Livingstone to Windhoek. Connect
with flight to London.
Day 14 Arrive in London.

Accommodation in a double sleeping cabin on the train and in a
double room in the Cape Town and Chobe Game Reserve. Single
beds available in the hotels. Extra night's accommodation in
Windhoek available.

Facilities in hotel: swimming pool, restaurant, bar, tennis.

Price per person includes air travel, all meals on the train, bed and
breakfast in the hotels, transfer to and from the airport. Not
included: travel insurance, visa, airport taxes, tips.

37 | ## *What will it be like in the future?*

Future simple (2): *(will)* for predictions

VOCABULARY AND LISTENING

знаки

1 Match the words in the box to the symbols for weather.

> fog cloud sun rain wind snow

fog

2 Which of these adjectives can you use to describe today's weather? *— туманный*

> cold cool dry foggy hot rainy sunny
> warm wet windy snowy

It's very hot today.

3 Look at the photos and say what the weather is like.

4 [cassette] Work in pairs. Look at the newspaper weather report below and listen to the radio weather forecast. Underline any information which is different from what you hear.

Worldwide forecast for midday tomorrow

Athens	c	12
Bangkok	c	30
Cairo	s	16
Geneva	c	4
Hong Kong	c	17
Istanbul	r	7
Kuala Lumpur	c	30
Lisbon	c	11
Madrid	r	7
Moscow	sn	−10
New York	s	0
Paris	sn	6
Prague	sn	−2
Rio	c	29
Rome	r	9
Tokyo	c	4
Warsaw	c	−8

GRAMMAR

— прогнозирование

> **Future simple (2): *(will)* for predictions**
> **You form the future simple with *will* + infinitive. You use the future simple to make predictions.**
> *It'll be sunny in New York tomorrow.*
> (= It will be sunny in New York tomorrow.)
> *It won't be rainy.* (= It will not be rainy.)
> *Will it be snowy in New York?*   Yes, it will.
> *Will there be rain in Geneva?*   No, there won't.

1 Work in pairs. Correct your answers to *Vocabulary and listening* activity 4.

In Geneva, it will be cloudy and ten degrees.

2 Look at the weather report again. Make comparisons between the weather in these cities.

1 Cairo – Tokyo
2 Athens – Rome
3 Warsaw – Moscow
4 Bangkok – Lisbon
5 Prague – Madrid
6 Istanbul – Athens

1 It'll be hotter in Cairo than in Tokyo.

3 Write a short forecast for tomorrow's weather in your town.

Tomorrow it'll be sunny with some clouds.

READING AND LISTENING

1 Look at these predictions about the weather. Do you think they are true for your country?

In twenty-five years:
– it'll be colder
– the sea level will be lower
– it'll be windier
– it'll be wetter
– there'll be more snow

2 Read *The temperature's rising* and find out if the predictions in 1 are true for Britain.

The temperature's rising

A government report says that in the next twenty-five years, Britain will get warmer and have higher sea levels. The weather will become more Mediterranean, and tourism will grow, but the Scottish ski industry will disappear because there will be little or no snow, and there'll be stronger winds. In the South there will be more sun, enough to produce wine, and in the North there will be more rain. It will be good for farmers, as crops will grow more quickly, and cattle and sheep will have warmer and wetter land in Scotland and northern England. But the higher sea level means that many towns, including London, will disappear under water.

People will only use heating in their homes for two or three months of the year, but they'll pay more for water. Snow at Christmas will become very rare. More people will die in the hotter summers, but the winters will be warmer as well.

3 Here are some more worldwide predictions. Which do you think will be true?

1 Temperatures will rise by two to six degrees Celsius in twenty-five years' time.
2 Ice at the North and South Pole will melt.
3 Whole countries will disappear underwater.
4 There won't be enough fresh water for everyone.
5 Fresh water will cost more.
6 Factory goods will cost more to produce.
7 The world economy will get worse.

4 🔲 Listen to a scientist talking about the predictions in 3. Put T if the predictions are true and F if they are false.

5 Work in pairs and check your answers to 4. Can you remember any details?

🔲 Listen again and check.

6 Work in groups of two or three. Make other predictions about the future.

There'll be more people in the world.

7 Tell other people about your predictions. Do they agree with you?

38 *Hamlet was written by Shakespeare*

Active and passive

SPEAKING AND VOCABULARY

1 Work in pairs. Are these sentences true or false?

Apple makes computers.
Lemons grow on trees.
Gustav Eiffel built the Eiffel Tower.
Marconi invented the radio.
Beethoven composed the Moonlight Sonata.
Leonardo da Vinci painted the Mona Lisa *(La Joconde)*.
Shakespeare wrote Hamlet.
Fleming discovered penicillin.

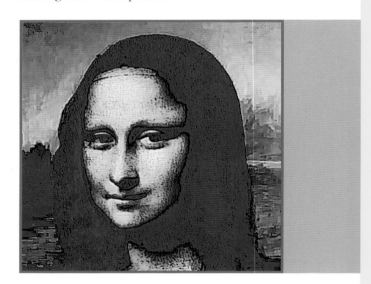

2 Complete these sentences with verbs from the box.
You may have to change the tense.

> make   grow   build   invent   compose   paint
> write   discover

1 Botticelli _____ *La Primavera*.
2 They _____ oranges in Spain.
3 Columbus _____ America.
4 Shah Jehan _____ the Taj Mahal.
5 Homer _____ the Odyssey.
6 They _____ Fiat cars in Italy.
7 The Chinese _____ gunpowder.
8 Tchaikovsky _____ the 1812 Symphony.

READING

1 Read *The round-the-world quiz* and choose the correct answer.

The round-the-world quiz

1 Coffee is grown in …
a Brazil   b England   c Sweden

2 Daewoo cars are made in …
a Switzerland   b Thailand   c Korea

3 Sony computers are made in …
a Japan   b the USA   c Germany

4 Tea is grown in …
a India   b France   c Canada

5 Tobacco is grown in …
a Norway   b Iceland   c the USA

6 Benetton clothes are made in …
a Italy   b France   c Malaysia

7 Roquefort cheese is made in …
a Germany   b Thailand   c France

8 The atom bomb was invented by …
a the Japanese   b the Americans   c the Chinese

9 *Guernica* was painted by …
a Picasso   b Turner   c Monet

10 The West Indies were discovered by …
a Scott of the Antarctic   b Christopher Columbus
c Marco Polo

11 The telephone was invented by …
a Bell   b Marconi   c Baird

12 *Romeo and Juliet* was written by …
a Ibsen   b Shakespeare   c Primo Levi

13 The Blue Mosque in Istanbul was built by …
a Sultan Ahmet I   b Ataturk
c Suleyman the Magnificent

14 *Yesterday* was composed by …
a Paul McCartney   b John Lennon   c Mick Jagger

15 The Pyramids were built by …
a the Pharaohs   b the Sultans   c the Council

2 Work in pairs. Check your answers to the quiz.

GRAMMAR

Active and passive	
Active	**Passive**
They **grow** *coffee in Brazil.*	*Coffee* **is grown** *in Brazil.*
The Americans **invented** *the atom bomb.*	*The atom bomb* **was invented** *by the Americans.* , при частие
You form the passive with the verb *to be* + past participle. You use *by* to say who or what is responsible for an action.	
Present simple	**Past simple**
Tea **is grown** *in India.*	*Guernica was painted* **by** *Picasso.*
Daewoo cars **are made** *in Korea.*	*The Pyramids were built* **by** *the Pharaohs.*

1 Look at the quiz again. Find examples of the passive. Are they in the present or past simple?

Sony computers are made in Japan.

2 Choose six sentences from the quiz and rewrite them.

Bell invented the telephone. They grow tobacco in the USA.

3 Rewrite the sentences in *Speaking and vocabulary* activity 2 in the passive.

La Primavera was painted by Botticelli.

LISTENING

1 📟 Listen to Frank and Sally doing *The round-the-world quiz*. Tick (✓) the correct answers. How many do they score? *Uол - количество очков.*

2 How many correct answers did you get?

WRITING AND SPEAKING

1 Work in pairs. Write a quiz about your country.

The Chapel Royal in the Wat Phrae Kaew was built by ....
a King Rama 1   b King Rama IV   c King Rama V

Make sure you include one correct ending and two incorrect ones.

2 Work with another pair. Do each other's quizzes.

39 | *She said it wasn't far*

Reported speech: statements

LISTENING AND READING

1 Decide where these sentences go in the conversation.

a ... it leaves at nine o'clock in the evening.
b And where's the next hostel?
c Where are you walking to?
d Yes, it's two kilometres.
e We'll stay for just one night.

CHRIS Good afternoon.

RECEPTIONIST Good afternoon. Can I help you?

CHRIS Have you got any beds for tonight?

RECEPTIONIST Yes, I think so. Sorry, but I've just started work at the hostel. How long would you like to stay?

CHRIS (1) _____ *Общее значение гостиница, турбаза*

RECEPTIONIST Yes, that's OK.

TONY Great!

RECEPTIONIST How old are you?

TONY We're both sixteen.

RECEPTIONIST One night's stay costs £6.50 each.

CHRIS Is it far from the hostel to the centre of Canbury?

RECEPTIONIST (2) _____ It takes an hour on foot.

TONY Is there a bus service?

RECEPTIONIST I think so. It takes about fifteen minutes. There's a bus every hour.

TONY When does the last bus leave the city centre?

RECEPTIONIST I think (3) _____ There's not much to do in the evening.

CHRIS We're very tired. We need an early night. What time does the hostel close in the morning?

RECEPTIONIST Er, at eleven am. (4) _____

CHRIS We're going to Oxton. Are you serving dinner tonight?

RECEPTIONIST Yes, we're serving dinner until eight o'clock. And breakfast starts at seven-thirty.

TONY (5) _____

RECEPTIONIST I'm not sure. I think it's Kingscombe, which is about ten kilometres away. I started work last Monday so I'm very new here.

2 🔲 Listen and check.

3 The receptionist gives Chris and Tony some wrong information. Read the brochure and underline the wrong information in the conversation.

> **Canbury Youth Hostel,**
> **Jackson Lane, Canbury**
> Tel 01789 3445 Fax 01789 3446
>
> **Facilities**
> 112 beds
> Open 24 hours, all year
> Car park
> No family rooms
> Games room
> Washing machines
> No smoking hostel
> Camping
>
> **Charges**
> Under 18 £6.15 Adult £9.10
>
> **Meal times**
> Breakfast 7am
> Dinner 6pm - 7pm
>
> **Travel Information**
> City centre three kilometres
> Bus service to city centre No.14
> (takes ten minutes) Last bus 8pm
>
> **Next hostels**
> Charlestown 8 kilometres
> Kingscombe 15 kilometres

4 🔊 Listen and number the next part of the conversation in the order you hear it.

TONY   And she said the last bus left at nine o'clock. But it leaves at eight o'clock. ☐

CHRIS   But, in fact, it's three kilometres. ☐

TONY   Yes, and she said it was two kilometres to the city centre. ☐

CHRIS   It's very strange. She said one night's stay cost £6.50, but it costs £6.15. ☐

GRAMMAR

косвенная речь

> **Reported speech: statements**
>
> **You report what people said by using *said (that)* + clause. If the tense of the verb in the direct statement is the present simple, the tense of the verb in the reported statement is the past simple. Pronouns also change.**
>
> Direct statement   Reported statement
>
> *The last bus **leaves** at nine o'clock,' he said.*   **He said that** *the last bus **left** at nine o'clock.*
>
> *It's two kilometres to the city centre,' he said.*   **He said** *it **was** two kilometres to the city centre.*
>
> Other tenses
>
> **Other tenses 'move back' in reported speech.**
>
> *I've just **started** work at the hostel,' he said.*
> **He said** *he **had** just **started** work at the hostel.*
> *We're going to Oxton,' he said.*
> **He said** *they **were going** to Oxton.*
> *We'll stay for just one night,' he said*
> **He said** *they **would stay** for just one night.*
> *I **started** work last Monday,' he said.*
> **He said** *he **had started** work last Monday.*

1 Write what the people actually said.

1 Tony said they were both sixteen.
2 The receptionist said it took an hour on foot.
3 She said there was a bus every hour.
4 She said there wasn't much to do in the evening.
5 Chris said they were very tired.
6 He said they needed an early night.

1 'We're both sixteen', said Tony.

2 Complete the rest of the conversation in *Listening and reading* activity 4.

CHRIS   And she said the bus _____ fifteen minutes. But in fact, it takes _____ minutes.

TONY   And she said the hostel _____ at eleven am, but it _____ open all day.

CHRIS   It seems that they _____ dinner from six to seven.

TONY   But she said they were serving until eight o'clock. And she also said breakfast _____ at seven-thirty....

CHRIS   ... when, in fact, it says here that breakfast _____ at seven.

TONY   And she said that Kingscombe _____ the next hostel, but it _____. It's Charlestown.

CHRIS   And finally she said that Kingscombe _____ ten kilometres away. But it _____ fifteen kilometres.

3 🔊 Listen and check.

VOCABULARY AND WRITING

1 Here are some new words from this lesson. Check you know what they mean.

> youth hostel   camping   no smoking   car park   facilities   charges   adult   travel

2 Complete the letter Chris and Tony wrote to the manager of Canbury Youth Hostel. Use as many of the words in the vocabulary box as possible.

> 9, King Street
> Shrewsbury
> SY2 6HJ
> 16 October, 1996
>
> The Manager
> Canbury Youth Hostel
> Jackson Lane
> Canbury
>
> Dear Sir,
> I'm writing to complain about the information your receptionist gave us when we stayed at the youth hostel. First of all, she said that one night's stay cost £6.50 when in fact, it costs £6.15. Then, she said that it was two ...

40 *Dear Jan ... Love Ruth*

Tense review

GRAMMAR

> **Tense review**
>
> **There are five tenses presented in *Reward* Elementary.**
>
> **Present simple**
> *I'm Polish. My name's Jan. What's your name?*
>
> **Present continuous**
> *You're staying with Mr and Mrs Hawkins. Mario is also living with them.*
>
> **Past simple**
> *A young man arrived at Brighton station. His name was Jan Polanski.*
>
> **Present perfect**
> *I've told him to go away. My parents haven't met many foreigners.*
>
> **Future simple**
> *I'll see you tomorrow. I'll miss you.*

1 Match the tenses 1 – 5 with their forms a – e below.

1 present simple
2 present continuous
3 past simple
4 present perfect
5 future simple

a most regular verbs: infinitive + *-ed*
b infinitive, or infinitive + *-s* for third person singular
c *am/is/are* + present participle
d *has/have* + past participle
e *will* + infinitive

2 Write the name of the tenses in 1 by their uses below.

a talking about recent events, such as a past action which has a result in the present
b saying what is happening now or around now
c making predictions
d talking about present customs and routines
e talking about finished actions in the past
f talking about experiences
g making decisions at the time of speaking

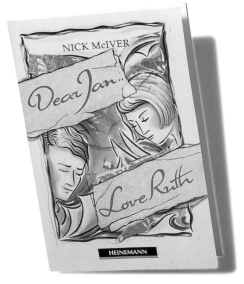

READING AND LISTENING

1 You're going to read a story called *Dear Jan ... Love Ruth*, by Nick McIver. Look at the cover of the book. What type of story do you think it is?

– a love story   – a detective story   – science fiction
– a mystery

2 Work in pairs. Part 1 of the story is called *The arrival*. Here are some words from part 1. Can you predict what happens?

Jan Polanski   Poland   language school   Brighton   stay family   disco   girls   boyfriends   stepped   foot   pretty name   Ruth   dance

3 Read part 1 and find out if you guessed correctly in 2.

The arrival

A young man arrived at Brighton station. His name was Jan Polanski and he came from Poland. He was in England for a course at an English language school. He took a taxi to the Modern Language Institute, went inside and met the director.

'Welcome to Brighton,' the director said. 'You're staying with the Hawkins family. Ah! Here's Mario. He's also living with them.'

'Hello, Jan,' said Mario.

That evening, after dinner Mario said, 'Would you like to come to a disco next Saturday?'

'Yes,' said Jan. 'Thanks very much.'

On Saturday Jan went to Mario's room. He was ill. 'I can't go to the disco tonight, Jan,' said Mario. 'But here's the address.'

Jan arrived at the disco at nine o'clock. He liked dancing, but most of the girls were with their boyfriends. Suddenly a girl stepped on his foot.

'Oh,' she said. 'I'm sorry.'

'That's all right,' said Jan. He looked at the girl. She was very pretty. 'Can I buy you a drink?' asked Jan. They went to the bar.

'You're not English, are you?' said the girl.

'No,' said Jan. 'I'm Polish. My name's Jan. What's your name?'

'Ruth,' she said. 'Ruth Clark.'

'Would you like to dance?' said Jan.

'Yes,' said Ruth.

4 What do you think happens next? Work in pairs and guess the answers to these questions about part 2.

1 Will they see each other again?
2 Where do they go?
3 Who does Ruth see?
4 What does she tell him to do?
5 Do Ruth and Jan like each other?

5 Read part 2 and find the answers to the questions in 4.

Jan and Ruth

The next day, Jan met Ruth and they walked by the sea. Then they went to a coffee bar. Suddenly a tall man came over to the table.

'Jan, can you go outside?' said Ruth.

Jan waited outside for about ten minutes. Then the man came out and walked away.

'Who was that?' asked Jan.

'That was Bill. He was my boyfriend. I've told him to go away. I don't like him any more.' She looked into Jan's eyes. 'Jan, I ... like you ... very much.'

Jan smiled. 'I like you very much too,' he said.

6 Work in pairs. Part 3 is called *Ruth's parents*. Here are some sentences from part 3 of the story. Decide who is speaking and to who.

1 'How do you do, Mr and Mrs Clark, ...'
2 'Sugar, but no milk.'
3 'These foreigners have strange ideas, ...'
4 'Your parents don't like me very much.'
5 'My parents haven't met many foreigners.'
6 'Well, he didn't speak English very well.'
7 'What's wrong with an English boyfriend?'
8 'But I don't like Bill any more, ...'
9 '... but I do like Jan. Maybe I love him.'

 Now listen and check.

7 Work in pairs. Part 4 is called *Going home*. Here are some phrases from part 4. What do you think happens next?

last day   goodbye   sad   come to Poland
living room   Bill was there   next morning
station   I'll miss you   I love you   train
started to leave

8 Turn to Communication activity 18 on page 103 and read part 4.

9 Work in pairs. Do you think the story has a happy or a sad ending? Talk about what will happen to Jan, Ruth ... and Bill.

If you'd like to know the ending, turn to Communication activity 24 on page 104.

VOCABULARY AND WRITING

1 Here are some words from the story. Check you remember what they mean.

language school   director   boyfriend   step
sea   coffee bar   foreigner   strange   miss
forget   love

2 Write a paragraph describing a different ending to the story.

In December, Ruth bought a train ticket to Poland...

Progress check 36–40

VOCABULARY

1 You put a preposition after many verbs.

listen to   agree with   decide to   laugh at

Match the verbs and the prepositions. Use your dictionary if necessary.

Verbs – apologise   belong   complain   go
hear   insist   pay   talk   think   worry

Prepositions – for   to   about   with   of   for
on   in

2 Here are some of the topics in *Reward* Elementary.

countries   jobs   family   furniture   entertainment
means of transport   shops   food and drink
clothes   sports

Try to think of two words which go with each topic.

3 Work in pairs and compare your answers to 2.

GRAMMAR

1 Make decisions about the following situations.

1 you've got nothing to eat
2 you're very tired
3 you don't feel well
4 you can't remember what a word means
5 you haven't got any money
6 you haven't spoken to your friend for a few days

1 I'll go shopping.

2 Make predictions about the following things.

1 tomorrow's weather
2 traffic in your town
3 the next World Cup
4 the next government of your country
5 your life in ten years' time
6 your English lessons

3 Rewrite these sentences in the passive.

1 Michelangelo painted the Sistine Chapel.
2 They grow cotton in Egypt.
3 They make Mercedes cars in Germany.
4 Hemingway wrote *The Old Man and the Sea*.
5 Verdi composed *Aida*.
6 William I built the Tower of London.

4 Rewrite these sentences in reported speech.

1 'I'm ill,' he said.
2 'It closes at seven,' she said.
3 'It leaves in five minutes,' she said.
4 'I work in an office,' he said.
5 'We live in London,' they said.
6 'She goes shopping on Saturday,' he said.

5 Write a sentence about yourself using each of these tenses.

1 present simple
2 present continuous
3 past simple
4 present perfect
5 future simple

SOUNDS

1 Say these words aloud. Is the underlined sound /ɔː/ or /ɔɪ/?

t<u>oy</u>   t<u>ore</u>   b<u>oy</u>   b<u>ore</u>   n<u>oise</u>   r<u>aw</u>   d<u>oor</u>   w<u>ar</u>   m<u>ore</u>

[cassette icon]   Listen and check. As you listen, say the words aloud.

сочинять

2 Underline the stressed syllable in these words.

invent   compose   discover   adult   travel   director
foreigner   forget

 Listen and check. As you listen, say the words aloud.

3 Listen and underline the words the speaker stresses.

> A young man arrived at Brighton station. His name was Jan Polanski. He came from Poland. He was in England for a course at an English language school. He took a taxi to the Modern Language Institute. He went inside and met the director.

Now read the passage aloud. Make sure you stress the same words.

SPEAKING

пересматривать
просматривать
сверять

You're going to revise the grammar and vocabulary you have learnt in *Reward* Elementary by playing *Reward Snakes and Ladders*. Work in groups of three or four and follow the instructions.

горки   лестница

Reward Snakes and Ladders

1 Look at the game board on Communication activity 1 on page 98.

2 Each player puts their counter on the square marked START and throws the dice. *игральная кость*

3 The first player to throw a six starts. *бросать*

4 Each player then throws the dice and moves his/her counter along the board according to the number thrown on the dice. As each player lands on a square, he/she has to answer the question on the square. (You can look back at the lesson to help you.) If a player answers the question correctly, he/she can remain on the square until their next turn. If a player answers a question incorrectly, he/she must go back 3 squares. The other players decide whether *будто* the answer given is right or wrong. If you land on a Progress Check square, the player on your left can ask you any question they like.

5 If you land on a ladder, you go up to the square shown and answer the question. If you land on a snake, you go down to the square shown and answer the question.

6 The winner is the first person to reach FINISH.

Communication activities

1 *Progress check 36–40*

Speaking

Now play
*Reward
Snakes and
Ladders.*

FINISH

Progress check 36 – 40 ?

40 Where's Jan staying? Have Ruth's parents met many foreigners?

39 When are the meals at the youth hostel?

38 Where is coffee grown? What do they grow in your country?

37 What will the weather be like tomorrow?

36 You need to get to London quickly. What will you do?

31 What must you or mustn't you do in your English lesson?

32 Have you ever been to London? Have you ever stayed in a hotel?

33 What's happened to Barry? What's happened to you today?

34 What's your advice for a perfect day out?

35 How did Beate pass her exam? How does Antonia play the guitar?

Progress check 31 – 35 ?

Progress check 26 – 30 ?

30 What's your favourite sport?

29 Is Thailand smaller than Britain? Is your country colder than Britain?

28 Name three parts of the body.

27 What's Joan's bag made of? When did she lose it?

26 Do you usually go shopping by yourself?

21 Why did Agatha Christie disappear?

22 When is New Year's Day? When is your birthday?

23 What's Harriet doing? What are you wearing?

24 What's Fiona going to do in the New Year? What's your New Year's resolution?

25 What would you like to eat? Do you like pasta?

Progress check 21 – 25 ?

Progress check 16 – 20 ?

20 Did Mary and Bill stay with friends? Where did you go on holiday?

19 What's Nick like? What are you like?

18 Where was Sting born? Where did Whitney Houston live as a child?

17 Do Jean and Tony need any fruit and vegetables? What do you need to buy?

16 Who was your first teacher? Who was your first friend?

11 What does Tanya Philips do in the evening? What do you usually do in the evenings?

12 How does Katie Francis get to work? How do you get to school/work?

13 Can cats swim? Can you drive?

14 Where's the bank? How do you get to the station?

15 Where are Janet and the kids staying? What are you doing at the moment?

Progress check 11 – 15 ?

Progress check 6 – 10 ?

10 Does Octavio like dancing? Do you like jazz?

9 How does Otto relax? How do you relax?

8 Where's the kitchen in your home? Have the Kapralovs got any chairs?

7 What's the time? What's the time in Britain?

6 What's Jenny's father's name? What's your mother's name?

1 Where is Marie from? Where are you from?

2 What's Michiko's job? What's your job?

3 Are you married? Is Greg Sheppard a doctor?

4 How many students are there in your school?

5 Where's your bag? Have you got a watch?

Progress check 1 – 5 ?

START

2 *Lesson 5*

Speaking and listening, activity 4

Look at this picture for 30 seconds.

Now turn back to page 11.

4 *Lesson 14*

Grammar and functions, activity 5

Student A: Look at the map below. Tell Student B where each shop and town facility is on your map.

Now listen to Student B and mark each shop and town facility he/she describes.

When you've finished, show your map to your partner. Have you marked the map correctly?

3 *Lesson 18*

Listening and speaking, activity 1

Student A: 🔲 Listen and find out:

- where Whitney Houston was born
- who she started singing with
- how many copies of her first two albums she sold
- when she appeared in *The Bodyguard*

Now turn back to page 43.

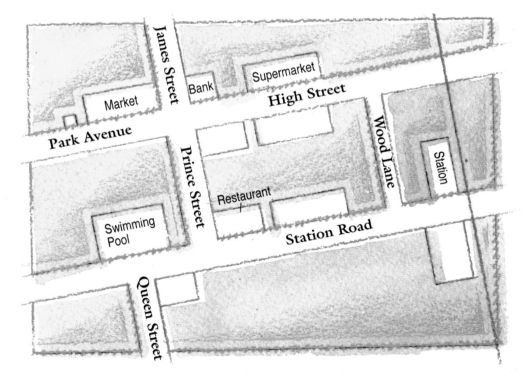

Now turn back to page 33.

5 *Lesson 22*

Speaking, activity 2

Pair A: Rewrite these facts as quiz questions.

1 The Kobe earthquake was in 1995.
2 Martin Luther King died in 1968.
3 Alexander Graham Bell invented the telephone.
4 Michelangelo was 88 years old when he died.
5 Joseph Stalin was born in Georgia.

1 When was the Kobe earthquake?

Now continue the quiz with Pair B. Ask and answer each other's questions.

6 *Lesson 21*

Grammar, activity 3

Write five statements about your past, three true and two false.

I was born in Belgium. I married Hercule Poirot.

Now work in pairs. Show your statements to your partner. Your partner must try to guess which are the false statements.

You didn't marry Hercule Poirot!

Now turn back to page 51.

7 *Lesson 23*

Functions and grammar, activity 4

Student A: Describe the photo to Student B. Now listen to Student B's description of a photo. Is the photo the same or different?

Now turn back to page 55.

8 *Lesson 21*

Vocabulary and reading, activity 5

Student A: Ask Student B these questions. Answer his/her questions in turn.

1 Why was Agatha Christie famous?
2 What was the final mystery?
3 When was she born?
4 Where did she live?
5 Who did she marry in 1914?

Now turn back to page 51.

9 *Lesson 36*

Reading and speaking, activity 3

Student B: You want to know more about the trip from the Cape to Victoria Falls. Ask Student A, who is the travel agent, the following questions:

Will there be a guided tour in Cape Town?
Will you have time to relax at the Victoria Falls?
Will the beds on the train be double or single beds?

With Student C, decide what you'll do.

10 *Lesson 2*

Vocabulary and sounds, activity 2

Look at the pictures for different jobs and check you know what the jobs are.

student   **receptionist**

secretary

teacher

farmer

Now turn back to page 4.

11 *Lesson 36*

Reading and speaking, activity 3

Student C: You want to know more about the trip from the Cape to Victoria Falls. Ask Student A, who is the travel agent, the following questions:

Will the flight to and from Cape Town be first class or tourist class?
Where will you stay in Johannesburg?
Will there be cabins with private bathrooms on the train?

With Student B, decide what you'll do.

12 *Lesson 14*

Grammar and functions, activity 5

Student B: Listen to Student A and mark the shops and town facilities he/she describes. Now tell Student A where each shop and town facility is on your map.

When you've finished, show your map to your partner. Have you marked the map correctly?

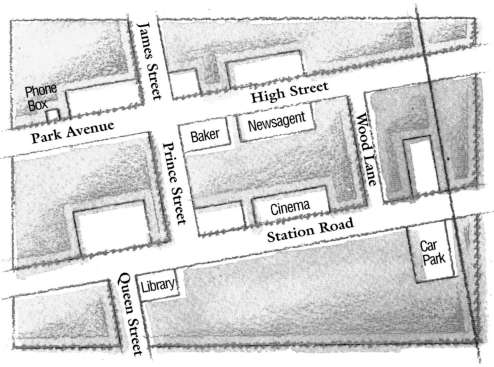

Now turn back to page 33.

13 *Lesson 23*

Vocabulary and listening, activity 1

Look at the photo for different items of clothing and check you know what the words mean.

jacket   sweater

skirt

socks   shoes   shorts

Now turn back to page 54.

14 *Lesson 6*

Vocabulary and sounds, activity 2

Look at the photo of Holly's family. Did you guess correctly in activity 1?

1 Andrew, brother
2 Antonia, Andrew's wife
3 David, grandfather
4 Philip, father
5 Steve, brother
6 Kate, grandmother
7 Jenny, mother
8 Holly, daughter

Now turn back to page 14.

15 *Lesson 21*

Vocabulary and reading, activity 5

Student B: Ask Student A these questions. Answer his/her questions in turn.

1 When did she write her first story?
2 What did she do in December 1926?
3 What did everyone think?
4 Where did her husband find her?
5 Who did she marry in 1930?

Now turn back to page 51.

16 *Lesson 28*

Functions and grammar, activity 3

Student A: Listen to Student B and say what he/she *should/shouldn't* do.

Now act out these situations with Student B. Listen to his/her advice.

  – you feel sick
  – you feel unhappy all the time
  – you are hungry
  – you don't like your job

Now turn back to page 66.

17 *Lesson 12*

Vocabulary and reading, activity 1

Look at the photos of different means of transport and check you know what the words mean.

train

bicycles

ferry

plane

underground

tram

car

18 *Lesson 40*

Reading and listening, activity 8. Part 4

Going home

It was the last day of the course at the Modern Language Institute and Jan was very sad. He said goodbye to Mario and his other friends and left the school. That night, Jan and Ruth went for a long walk in Brighton.

'I love you, Ruth,' Jan said.

'I love you too, Jan.'

'I'm going home tomorrow. But why don't you come to Poland at Christmas?' said Jan.

'Yes,' said Ruth. 'I'd love to.'

Jan suddenly laughed. 'I'm going to see you again!'

Ruth got home at eleven o'clock that evening. She went into the house and her mother met her in the hall.

'You've got a visitor, Ruth,' she said.

Ruth went into the living room. Bill was there.

The next morning, Ruth went to Brighton station with Jan.

Jan said, 'It's September now. And you're coming to Poland in December.'

'I know,' said Ruth. 'But I'll miss you.'

At that moment, the train started to leave.

'Goodbye, Ruth,' said Jan. 'I love you.'

'Goodbye, Jan. Write to me.'

'Yes, of course.'

Ruth walked away from the station. She went down to the sea and thought about Jan.

Now turn back to page 95.

19 *Lesson 36*

Reading and speaking, activity 3

Student A: You're a travel agent. Student B and C want to know more about the trip from the Cape to Victoria Falls. Read the information below and answer their questions.

- there will be a personal guided tour in Cape Town or a large group tour
- they will be able to stay in a hotel or on the train while in Johannesburg
- there will be time to relax at the Victoria Falls or they will be able to go to the game reserve
- the flight to and from Cape Town will be first class or tourist class
- the beds on the train will be double beds or two single beds
- there will be cabins on the train with or without bathrooms

Ask them to make a decision.

20 *Lesson 4*

Sounds and vocabulary, activity 2

Check your answers.

3, 13, 30, 4, 14, 40, 5, 15, 50, 6, 16, 60, 7, 17, 70, 8, 18, 80, 9, 19, 90 100

Now turn back to page 8.

21 *Lesson 28*

Functions and grammar, activity 3

Student B: Act out these situations with Student A and listen to his/her advice.

- you don't get enough exercise

- you feel tired all the time

- you are thirsty

- you have got a headache

Now listen to Student A and say what he/she *should/shouldn't* do.

Now turn back to page 66.

22 *Lesson 22*

Speaking, activity 2

Pair B: Rewrite these facts as quiz questions.

1 The Mexico City earthquake was in 1984.
2 Yuri Gagarin was the first man in space.
3 The first American walked on the moon in 1969.
4 King Henry VIII of England and Wales had six wives.
5 Sigmund Freud was born in Vienna.

1 When was the Mexico City earthquake?

Now continue the quiz with Pair A. Ask and answer each other's questions.

23 *Lesson 18*

Listening and speaking, activity 1

Student B: Listen and find out:

- when Whitney Houston was born
- when she started singing
- when she had her first hit
- who she appeared with in *The Bodyguard*
- when she had a hit with *I will always love you*

Now turn back to page 43.

24 *Lesson 40*

Reading and listening, activity 9. Part 5

The end

Jan wrote several letters to Ruth. But every time Mrs Clark found the letters she burnt them. Ruth was very sad. She thought, Jan doesn't love me any more. He's forgotten about me.

Bill was very kind to Ruth at this time. At the end of November, Ruth went to a party with Bill.

Now turn back to page 95.

25 *Lesson 23*

Reading and speaking, activity 3

Add up your scores using the following table. Then look at the profiles below to find out what your clothes say about you.

1 a 2   b 3   c 1
2 a 2   b 3   c 1
3 a 2   b 1   c 3
4 a 2   b 1   c 3
5 a 3   b 2   c 1
6 a 3   b 2   c 1
7 a 3   b 2   c 1
8 a 2   b 1   c 3
9 a 1   b 3   c 2
10 a 1   b 2   c 3

21 – 30 points. You like to wear exactly what you want. Sometimes this may get you into trouble.

11 – 20 points. You are quite casual. Sometimes you don't wear the right clothes for the situation.

1 – 10 points. You're very careful to wear the right clothes for the right situation.

26 *Lesson 13*

Vocabulary and reading, activity 4

11 – 15 points.   Excellent! You're an all-rounder!

6 – 10 points.   Good. You can do many different things.

1– 5 points.   You're not an all-rounder! You like to specialise!

Now turn back to page 30.

27 *Lesson 23*

Functions and grammar, activity 4

Student B: Listen to Student A's description of a photo. Now describe your photo to Student A. Is the photo the same or different?

Now turn back to page 55.

28 *Lesson 30*

Vocabulary and listening, activity 1

Look at the photos of different sports and check you know what they are.

Grammar review

CONTENTS

Present simple

Form

You use the contracted form in spoken and informal written English.

Be

Affirmative	Negative
I'm (I am)	I'm not (am not)
you	you
we 're (are)	we aren't (are not)
they	they
he	he
she 's (is)	she isn't (is not)
it	it

Questions	Short answers
Am I?	Yes, I am.
	No, I'm not.
Are you/we/they?	Yes, you/we/they are.
	No, you/we/they're not.
Is he/she/it?	Yes, he/she/it is.
	No, he/she/it isn't.

Have

Affirmative	Negative
I	I
you have	you haven't (have not)
we	we
they	they
he	he
she has	she hasn't (has not)
it	it

Questions	Short answers
Have I/you/we/they?	Yes, I/you/we/they have.
	No, I/you/we/they haven't.
Has he/she/it?	Yes, he/she/it has.
	No, he/she/it hasn't.

Regular verbs

Affirmative		Negative	
I		I	
you	work	you	don't (do not) work
we		we	
they		they	
he		he	
she	works	she	doesn't (does not) work
it		it	

Questions	Short answers
Do I/you/we/they work?	Yes, I/you/we/they do.
	No, I/you/we/they don't (do not).
Does he/she/it work?	Yes, he/she/it does.
	No, he/she/it doesn't (does not).

Question words with *is/are*
What's your name?   Where are your parents?

Question words with *does/do*
Where does he live?   What do you do?

Present simple: third person singular

You add *-s* to most verbs.
takes, gets
You add *-es* to *do, go* and verbs which end in
-ch, -ss, -sh and *-x.*
does, goes, watches, finishes
You drop the *-y* and add *-ies* to verbs ending in *-y.*
carries, tries

Use
You use the present simple:

- to talk about customs. (See Lesson 7)
  In Spain people have dinner at ten or eleven in the evening.
  In Britain people leave work at five in the afternoon.

- to talk about habits and routines. (See Lesson 9)
  I go running every day.
  We see friends at the weekend.

- to say how often you do things. (See Lesson 11)
  I always get up at seven o'clock.
  I sometimes do the shopping in the evening.

- to describe something that is true for a long time. (See Lesson 23)
  He wears glasses.

Present continuous

Form
You form the present continuous with *be* + present participle (*-ing*). You use the contracted form in spoken and informal written English.

Affirmative		Negative	
I'm (am) working		I'm not (am not) working	
you		you	
we	're (are) working	we	aren't (are not) working
they		they	
he		he	
she	's (is) working	she	isn't (is not) working
it		it	

Questions	Short answers
Am I working?	Yes, I am.
	No, I'm not.
Are you/we/they working?	Yes, you/we/they are.
	No, you/we/they aren't.
Is he/she/it working?	Yes, he/she/it is.
	No, he/she/it isn't.

Question words
What are you doing?   Why are you laughing?

Present participle (*-ing*) endings

You form the present participle of most verbs by adding *-ing*:
go – going   visit – visiting

You drop the *-e* and add *-ing* to verbs ending in *-e.*
make – making   have – having

You double the final consonant of verbs of one syllable ending in a vowel and a consonant and add *-ing.*
get – getting   shop – shopping

You add *-ing* to verbs ending in a vowel and *-y* or *-w.*
draw – drawing   play – playing

You don't usually use these verbs in the continuous form.
believe feel hate hear know like love see smell sound taste think understand want

Use
You use the present continuous:

- to describe something that is happening now or around now. (See Lessons 15 and 23)
  We're flying at 10,000 metres.
  She's wearing a yellow dress.

- to talk about future intentions or plans which are fairly certain. (See Lesson 24)
  We're going to see Mary on Saturday.
  He's coming round this evening.

Past simple

Form
You use the contracted form in spoken and informal written English.

Be

Affirmative	Negative
I	I
he was	he wasn't (was not)
she	she
it	it
you	you
we were	we weren't (were not)
they	they

Have

Affirmative	Negative
I	I
you	you
we	we
they had	they didn't (did not) have
he	he
she	she
it	it

Regular verbs

Affirmative	Negative
I	I
you	you
we	we
they worked	they didn't (did not) work
he	he
she	she
it	it

Questions	Short answers
Did I/you/we/they work?	Yes, I/you/we/they did.
he/she/it	he/she/it
	No, I/you/we/they didn't.
	he/she/it

Question words
What did you do yesterday?　　*Why did you leave?*

Past simple endings

You add *-ed* to most regular verbs.
walk – walked　　watch – watched

You add *-d* to verbs ending in *-e*.
close – closed　　continue – continued

You double the consonant and add *-ed* to verbs of one syllable ending in a vowel and a consonant.
stop – stopped　　plan – planned

You drop the *-y* and add *-ied* to verbs ending in *-y*.
study – studied　　try – tried

You add *-ed* to verbs ending in a vowel + *-y*.
play – played　　annoy – annoyed

Irregular verbs
There are many verbs which have an irregular past simple. For a list of the irregular verbs which appear in **Reward Elementary**, see page 114.

Pronunciation of past simple endings
/t/ *finished, liked, walked*
/d/ *continued, lived, stayed*
/ɪd/ *decided, started, visited*

Expressions of past time
(See Lesson 22)

yesterday morning/afternoon/evening
last Saturday/week/month/year
two weeks ago/six months ago

Use
You use the past simple:

● to talk about an action or event in the past that is finished. (See Lessons 16, 18, 20 and 22)
What were you like as a child?
I started learning English last year.
Did they go to France last year?

Future simple (*will*)

Form
You form the future simple with *will* + infinitive. You use the contracted form in spoken and informal written English.

Affirmative	Negative
I	I
you	you
we	we
they 'll (will) work	they won't (will not) work
he	he
she	she
it	it

Questions	Short answers
Will I/you/we/they work?	Yes, I/you/we/they will.
he/she/it/	he/she/it/
	No, I/you/we/they won't.
	he/she/it/

Question words
What will you do?　　*Where will you go?*

Use

You use the future simple:

● to talk about a decision you make at the moment of speaking. (See Lessons 26 and 36)
I'll have the blue T-shirt.
I think I'll go out tonight.
I'll call back later.

● to make a prediction or express an opinion about the future. (See Lesson 37)
There'll be more and more people in the world.
I think it'll be hot tomorrow.

Present perfect simple

Form

You form the present perfect simple with *has/have* + past participle. You use the contracted form in spoken and informal written English.

Affirmative		Negative	
I		I	
you	've (have) worked	you	haven't (have not) worked
we		we	
they		they	
he		he	
she	's (has) worked	she	hasn't (has not) worked
it		it	

Questions	Short answers
Have I/you/we/they worked?	Yes, I/you/we/they have.
	No, I/you/we/they haven't.
Has he/she/it worked?	Yes, he/she/it has.
	No, he/she/it hasn't.

Past participles

All regular and some irregular verbs have past participles which are the same as their past simple form.
Regular: *move – moved, finish – finished, visit – visited*
Irregular: *leave – left, find – found, buy – bought*

Some irregular verbs have past participles which are not the same as the past simple form.
go – went – gone   be – was/were – been
drink – drank – drunk   ring – rang – rung

For a list of the past participles of the irregular verbs which appear in **Reward Elementary**, see page 114.

Been and gone

He's been to America. = He's been there and he's back here now.
He's gone to America. = He's still there.

Use

You use the present perfect simple:

● to talk about past experiences. You often use it with *ever* and *never*. (See Lesson 32)
Have you ever stayed in a hotel? (=Do you have the experience of staying in a hotel?)
Yes, I have. (=Yes, I have stayed in a hotel at some point, but it's not important when.)
No, I've never stayed in a hotel.

Remember that if you ask for and give more information about these experiences, actions or states, such as *when, how, why* and *how long*, you use the past simple.
When did you stay in a hotel?   When I was in France last year.

● to talk about recent events, such as a past action which has a result in the present. You often use it to describe a change. (See Lesson 33)
He's hurt his back.
He's lost his wallet.

You use *just* if the action is very recent.
He's just lost his wallet.

You use *yet* in questions and negatives to talk about an action which is expected. You usually put it at the end of the sentence.
Have you been to the doctor yet?

Remember to use the past simple to say when the action happened.
He hurt his back this morning.

Questions

You can form questions in two ways:

● without a question word. (See Lesson 3)
Are you British?
Was he born in Italy?
Do you have any brothers?
Did you get up late this morning?

● with a question word such as *who, what, where, when, how* and *why*. (See Lesson 9)
What's his job?
How old is he?
What do you do to relax?
Where were you born?

You can put a noun after *what* and *which*.
What time is it? *Which road will you take?*

You can put an adjective or an adverb after *how*.
How much is it? *How long does it take by car?*
How fast can you drive?

You can use *who, what* or *which* as pronouns to ask about the subject of the sentence. You don't use *do* or *did*.
What's your first name? *Who was Agatha Christie?*

You can use *who, what* or *which* as pronouns to ask about the object of the sentence. You use *do* or *did*.
What did Agatha Christie do? *Who did she marry?*

You can form more indirect, polite questions with one of the following question phrases.
Can I help you?
Could I have some water, please?
Would you like a regular or a large Coke?

Imperatives

The imperative has exactly the same form as the infinitive (without *to*) and does not usually have a subject. You use the imperative:

- to give directions. (See Lesson 14)
  Go along East Street.
  Turn left into Prince Street.
- to give instructions and advice. (See Lesson 34)
  Come in.
  Sit down.
  Check the weather forecast before you go.

You use *don't* + imperative to give a negative instruction.
Don't take too much to carry.

Verb patterns

There are several possible patterns after certain verbs which involve -*ing* form verbs and infinitive constructions with or without *to*.

-*ing* form verbs

You can put an -*ing* form verb after certain verbs. (See Lesson 10)
I like playing football on the beach.
Peter hates travelling by plane.

Remember that *would like to do something* refers to an activity at a specific time in the future.
I'd like to go to the cinema next Saturday.

When you *like doing something*, this is something you enjoy all the time.
I like going to the cinema. I go most weekends.

to + infinitive

You can put *to* + infinitive after many verbs.
Here are some of them:
decide go have hope learn like need want
He decided to go to Spain for a holiday.

Use

You use *to* + infinitive (the infinitive of purpose):

- to say why people do things. (See Lesson 34)
  You go to a chemist to buy suncream.
  You go to the bus stop to catch a bus.
- to describe the purpose of something. (See Lesson 34)
  You use ice to keep things cold.

Have got

Form

You use the contracted form in spoken and informal written English.

Affirmative		Negative	
I		I	
you	've (have) got	you	haven't (have not) got
we		we	
they		they	
he		he	
she	's (has) got	she	hasn't (has not) got
it		it	

Questions	Short answers
Have I/you/we/they got?	Yes, I/you/we/they have.
	No, I/you/we/they haven't.
Has he/she/it got?	Yes, he/she/it has.
	No, he/she/it hasn't.

Use

You use *have got* to talk about possession.
He's got a new car.
Have got means the same as *have*. You use it in spoken and informal written English.
She's got a house in London. (= She has a house in London.)

Going to

You use *going to* + infinitive:

- to talk about future intentions or plans which are fairly certain. (See Lesson 24)
  I'm going to see my friends more often.
- to talk about something that we can see now is sure to happen in the future. (See Lesson 24)
  She's going to have a baby.

Modal verbs

The following verbs are modal verbs.
can   could   must   should   will   would   needn't

Form

Modal verbs:

● have the same form for all persons.
I must go.   He must be quiet.

● don't take the auxiliary *do* in questions and negatives.
Can you use a computer?
You mustn't be late for the meeting.

● take an infinitive without *to*.
I can type.
You should see a doctor.

Use

You use *can*:

● to talk about general ability, something you are able to do on most occasions. (See Lesson 13)
I can play the piano.
I can drive a car.

● to ask for permission. (See Lesson 26)
Can I try this on?

● to say if you're allowed to do something (See Lesson 31)
You can go to London on Saturday.

You can also use *could*. *Can* is a little less formal than *could*.

You use *could*:

● to ask for something politely.
Could I have some water, please?

● to ask people to do things
Could you tell me your name?

● to ask for permission
Could I try this on?

You use *must*:

● to talk about something you're strongly advised to do or are obliged to do. (See Lesson 31)
You really must stop smoking.
It's late. I must go now.
You must stop at a red light.

You often use it to talk about safety instructions.
You must use your seatbelt.

You use *mustn't*:

● to talk about something you're strongly advised not to do. (See Lesson 31)
You mustn't point at people.

● to talk about something you're not allowed to do. (See Lesson 31)
You mustn't drive through a red light.

You can also use *can't*.
You can't drive through a red light.

Remember that you don't usually use *must* in questions.

You use *should*:

● to give less strong advice. It can also express a mild obligation or the opinion of the speaker. (See Lesson 28)
You should go to bed.
You shouldn't go to work.

For uses of *will* see Future simple (*will*).

You use *would like* + noun or *would like to* + infinitive:

● to offer or request something politely. (See Lesson 25)
Would you like a drink?
What would you like to drink?
I'd like a burger, please.
I'd like to go to the cinema tonight.

You can also use *like* to say what you like all the time.
I like Coke. (= always)
I'd like a Coke. (= now)

You use *needn't* to talk about something it isn't necessary to do.
You needn't do your homework tonight.

Pronouns

Subject	**Object**	**Possessive**
(See Lesson 10)	(See Lesson 10)	(See Lesson 27)
I	me	mine
you	you	yours
he	him	his
she	her	hers
it	it	its
we	us	ours
they	them	theirs

Articles

There are many rules for the use of articles. Here are some of the most useful. (See Lessons 2 and 12)
You use the indefinite article (*a/an*):

● to talk about something for the first time.
She works in an office in Paris.
I get a train to work.

● with jobs.
She's a marketing consultant.
He's a ticket inspector.

- with certain expressions of quantity.
  I go to the cinema once or twice a month.
  There are several trains a day.

You use *an* for nouns which begin with a vowel.
an accountant, an apple

You use the definite article (*the*):

- to talk about something again.
  The office is near the Gare du Nord.
  I get the eight o'clock train.

- when there is only one.
  The Channel Tunnel.
  The Eurostar service.

Before vowels you pronounce *the* /ði:/.
You don't use any article:

- with certain expressions.
  by train   by plane   at work   at home

- with most countries, meals, languages.
  She often goes to France.
  She lives in Britain.
  Let's have lunch.
  I speak Italian.

Plurals

You form the plural of most nouns with *-s*.
(See Lessons 4 and 6)
bag – bags, book – books, key – keys

For nouns which end in *-y*, you drop *-y* and add *-ies*.
diary – diaries, baby – babies

You add *-es* to nouns which end in *-o, -ch, -ss, -sh* and *-x*.
watch – watches, glass – glasses

There are some irregular plurals.
man – men, woman – women, child – children

Possessives

Possessive *'s*

You add *'s* to singular nouns to show possession.
(See Lesson 6)
John's mother.   His teacher's book.

You add *s'* to regular plural nouns.
My parents' names are Georges and Paulette.
The boys' room.

You add *'s* to irregular plural nouns.
Their children's names are Pierre and Thierry.
The men's room.

Possessive adjectives

You can find the main uses for possessive adjectives in Lesson 2.

Form						
I	you	he	she	it	we	they
my	your	his	her	its	our	their

Whose

You use *whose* to ask who something belongs to.
(See Lesson 27)
Whose bag is this?
Whose are these shoes?

Expressions of quantity

Countable and uncountable nouns

Countable nouns have both a singular and a plural form.
(See Lesson 17)
a banana – bananas, a tomato – tomatoes

Uncountable nouns do not usually have a plural form.
water, juice, wine

If you talk about different kinds of uncountable nouns, they become countable.
Beaujolais and Bordeaux are both French wines.

Some and *any*

You usually use *some* with plural and uncountable nouns in affirmative sentences when you are not interested in the exact number. (See Lessons 8 and 17)
We need some fruit and vegetables.

You usually use *any* with plural and uncountable nouns in negative sentences and questions. (See Lessons 8 and 17)
We haven't got any carrots.
Have we got any milk?

You use *some* in questions when you ask for, offer or suggest something.
How about some oranges?

Adjectives

Position of adjectives

You can put an adjective in two positions. (See Lesson 4)

- after the verb *to be*.
  The book is very interesting.

- before a noun.
  It's an interesting book.

Comparative and superlative adjectives

Form

You add *-er* to most adjectives for the comparative form, and *-est* for the superlative form. (See Lessons 29 and 30)

cold – colder – coldest   *cheap – cheaper – cheapest*

You add *-r* to adjectives ending in *-e* for the comparative form, and *-st* for the superlative form.

large – larger – largest   *fine – finer – finest*

You add *-ier* to adjectives ending in *-y* for the comparative form, and *-iest* for the superlative form.

happy – happier – happiest   *friendly – friendlier – friendliest*

You double the final consonant and add *-er* or *-est* to adjectives of one syllable ending in a vowel and a consonant.

hot – hotter – hottest   *thin – thinner – thinnest*

You use *more* for the comparative form and *most* for the superlative form of longer adjectives.

expensive – more expensive – most expensive
important – more important – most important

Some adjectives have irregular comparative and superlative forms.

good – better – best   *bad – worse – worst*

With the superlative form you usually use *the* before the adjective in its superlative form.

James is the tallest person in the room.

You use a comparative + *than* when you compare two things which are different.

Thailand is bigger than Britain.

Adverbs

Formation of adverbs

You use an adverb to describe a verb. (See Lesson 35)

She speaks English fluently.
He drives carelessly.

You form an adverb by adding *-ly* to the adjective.

fluent – fluently   careless – carelessly

If the adjective ends in *-y*, you drop the *-y* and add *-ily*.

happy – happily   easy – easily

Some adverbs have the same form as the adjective they come from.

late, early, hard

The adverb from the adjective *good* is *well*.

She's a good writer.   *She writes well.*

Position of adverbs of frequency

You usually put adverbs of frequency before the verb. (See Lesson 11)

I always get up at seven o'clock.
I often have a drink with friends.

But you put them after the verb *to be*.

I'm never late for work.

Prepositions of time and place

in, at, on, to

You use *in*:
- with seasons and months of the year.
  in winter, in September, in March
- with places. (See Lesson 5)
  in the classroom, in the photograph, in Greece
- with times of the day. (See Lesson 7)
  in the morning, in the afternoon

You use *at*:
- with certain expressions. (See Lesson 11)
  at school, at home, at work
- with times of the day. (See Lesson 11)
  at night, at seven o'clock, at the weekend

You use *on*:
- with days and dates. (See Lesson 11)
  on Sunday, on Monday morning, on 15th June

You use *to*:
- with places.
  Helena goes to London every month.

You use *from ... to*:
- to express how long something lasts. (See Lesson 11)
  The shop is open from seven to nine o'clock.

Present simple passive

You form the present simple passive with *am/is/are* + past participle. (See Lesson 38)
Fiat cars are made in Italy.

Use

You use the passive to focus on the object of the sentence.
This palace was built by Palladio in 1635.

Reported speech

Statements

You report what people said by using *said that* + clause. Notice how the tense of the verb in the direct statement moves one tense back in the reported statement.

Direct statement	**Reported statement**
'The film finishes at ten o'clock,' she said.	*She said the film finished at ten o'clock.*
'We're going on holiday next year,' she said	*She said they were going on holiday next year.*
'We'll catch the train to Paris,' he said.	*He said they would catch the train to Paris.*
'I watched the television all evening,' he said.	*He said he had watched the television all evening.*

Irregular Verbs

Verbs with the same infinitive, past simple and past participle

cost	cost	cost
cut	cut	cut
hit	hit	hit
let	let	let
put	put	put
read /ri:d/	read /red/	read /red/
set	set	set
shut	shut	shut

Verbs with the same past simple and past participle but a different infinitive

bring	brought	brought
build	built	built
burn	burnt/burned	burnt/burned
buy	bought	bought
catch	caught	caught
feel	felt	felt
find	found	found
get	got	got
have	had	had
hear	heard	heard
hold	held	held
keep	kept	kept
learn	learnt/learned	learnt/learned
leave	left	left
lend	lent	lent
light	lit/lighted	lit/lighted
lose	lost	lost
make	made	made
mean	meant	meant
meet	met	met
pay	paid	paid
say	said	said
sell	sold	sold
send	sent	sent
sit	sat	sat
sleep	slept	slept
smell	smelt/smelled	smelt/smelled
spell	spelt/spelled	spelt/spelled
spend	spent	spent
stand	stood	stood
teach	taught	taught
understand	understood	understood
win	won	won

Verbs with same infinitive and past participle but a different past simple

become	became	become
come	came	come
run	ran	run

Verbs with a different infinitive, past simple and past participle

be	was/were	been
begin	began	begun
break	broke	broken
choose	chose	chosen
do	did	done
draw	drew	drawn
drink	drank	drunk
drive	drove	driven
eat	ate	eaten
fall	fell	fallen
fly	flew	flown
forget	forgot	forgotten
give	gave	given
go	went	gone
grow	grew	grown
know	knew	known
lie	lay	lain
ring	rang	rung
rise	rose	risen
see	saw	seen
show	showed	shown
sing	sang	sung
speak	spoke	spoken
swim	swam	swum
take	took	taken
throw	threw	thrown
wake	woke	woken
wear	wore	worn
write	wrote	written

Pronunciation guide

/ɑː/	park	/b/	buy
/æ/	hat	/d/	day
/aɪ/	my	/f/	free
/aʊ/	how	/g/	give
/e/	ten	/h/	house
/eɪ/	bay	/j/	you
/eə/	there	/k/	cat
/ɪ/	sit	/l/	look
/iː/	me	/m/	mean
/ɪə/	beer	/n/	nice
/ɒ/	what	/p/	paper
/əʊ/	no	/r/	rain
/ɔː/	more	/s/	sad
/ɔɪ/	toy	/t/	time
/ʊ/	took	/v/	verb
/uː/	soon	/w/	wine
/ʊə/	tour	/z/	zoo
/ɜː/	sir	/ʃ/	shirt
/ʌ/	sun	/ʒ/	leisure
/ə/	better	/ŋ/	sing
		/tʃ/	church
		/θ/	thank
		/ð/	then
		/dʒ/	jacket

Tapescripts

Lesson 5   Listening and vocabulary, activity 2

Conversation 1
MAN Where's my pen?
WOMAN It's on the table, near your book.
MAN Oh, I see. Thanks.

Conversation 2
MAN Have you got a mobile phone?
WOMAN Yes, I have. It's in my bag. Here you are.
MAN Thanks.

Conversation 3
WOMAN Where's my bag?
MAN What colour is it?
WOMAN It's blue.
MAN It's under your chair.
WOMAN Oh, yes. Thank you.

Lesson 5   Speaking and listening, activity 2

Q Tell me about your personal possessions, Steve. What sort of things have you got?
STEVE Well, I've got a personal stereo, I suppose that's very typical these days. And I've got a computer, like everyone else.
Q Have you got a mobile phone?
STEVE No, I haven't. I don't need one at the moment. I'm still at school.
Q And have you got a bicycle?
STEVE Yes, I have. Oh, and I've got a television and a radio in my bedroom.
Q And have you got a video too?
STEVE No, I haven't. My parents have got a video, but it's downstairs.
Q And what about a watch?
STEVE Yes, I've got a watch. Oh, I'm late for school! Bye!
Q Thanks, Steve.

Lesson 7   Reading and listening, activity 3

Q So, Tony, you're Australian, right?
TONY That's right.
Q And where do you come from in Australia?
TONY From Sydney.
Q Sydney! I've heard it's very beautiful there.
TONY I think it's *very* beautiful, but it *is* my hometown.
Q Tell me about the daily routine in Australia. What time do you get up?
TONY During the week, we get up at seven in the morning.
Q What time do children start school in the morning?
TONY It's usually about nine o'clock.
Q Nine o'clock. That's later than many countries.
TONY Yes, it is.
Q And when do they finish school?
TONY At about three in the afternoon.
Q And when do people go to work in the morning?
TONY Well, we start work at nine o'clock, so we go to work at seven-thirty or eight o'clock.
Q So they start work at nine o'clock, you say?
TONY That's right. Nine o'clock.
Q And when do you have lunch?
TONY Well, we have lunch at one o'clock. We stop work and have a sandwich usually, but our main meal is dinner, in the evening.
Q And what time do you stop work?
TONY We stop work at five in the afternoon. Actually, we leave work at five in the afternoon. We probably stop work earlier!
Q And when do you have dinner?
TONY At seven o'clock in the evening, usually. We eat outside in the garden most of the year.
Q And when do you go to bed?
TONY We go to bed at eleven or twelve at night.
Q And do you work on Saturdays and Sundays?
TONY No, we don't work at the weekend.

Lesson 8   Listening and writing, activity 1

Q So Geoff, this is your home!
GEOFF That's right. Do you like it?
Q I like it very much. It's a very nice boat. It's so quiet here on the river.
GEOFF It's noisy in the summer with the tourist boats, but in winter it's perfect.
Q How big is it?
GEOFF Well, it's a special type of boat for the canals, which are very narrow in Britain. It's called a narrow boat and it's ten metres long and about two metres wide.
Q Ten metres! Is it difficult to drive it along the canals?
GEOFF At first it's difficult, but after a while, with practice, it's quite easy.
Q But I suppose with ten metres, you have a lot of space.
GEOFF Yes, well, we're in the living room, and there's a kitchen, a bathroom and two bedrooms through there.
Q How many people live in the boat with you?
GEOFF My wife and our baby daughter, three people in all.
Q Three of you, I see. And what sort of furniture do you have?
GEOFF Well, there's a fridge and a cooker, several armchairs, a television and a shower, but there isn't room for a bath and a dishwasher. But it's quite comfortable.
Q And what's the most important item for you?
GEOFF I suppose it's my computer. Yes, it's my computer.

Lesson 9   Listening and speaking, activity 2

Q So what do you do in your free time, Helen?
HELEN Well, I like to relax with a good novel.
Q Really? You like reading then?
HELEN Yes, I do. I like newspapers and magazines but for me a good novel is the best.
Q How about you Chris?
CHRIS Well, I haven't got much time for reading. I see my friends a lot and we go to a club every Friday and Saturday.
Q And what about during the rest of the week?
CHRIS Well, I stay at home and watch television.
Q Do you watch television, Helen?
HELEN No, not very much. There's a television in the sitting room, but I only watch the news. Oh, there is one programme I like – it's called *Sueños*.
Q *Sueños?* What's that?
HELEN Well, it's a very good programme on how to learn Spanish. It's on BBC. I learn languages in my free time. Last year I learned French, and next year I'd like to do Russian.
Q That's fascinating. Thank you.

Lesson 10   Listening, activity 1

A Do you like rock music?
B No, I don't. I hate it.
A What type of music do you like? Do you like jazz?
B Yes, I do. I love it.
A Who's your favourite musician?
B Miles Davis. He's great.
A I don't like him very much. I don't like jazz.
B Oh, I like it very much. What about you? Who's your favourite singer?
A I like classical music. My favourite singer is Pavarotti. Do you like classical music?
B It's all right.

Lesson 10 Listening and writing, activity 2

JOHN Hi, I'm John, I live in Brighton in England. I'm a student at university here. I like sport, football, tennis and skiing. And my favourite music is rock music. I like the *Rolling Stones*.

KATE Hello, my name's Kate and I live in Edinburgh, in Scotland. I like winter sports like skiing and ice skating, because it gets quite cold here in Scotland. In fact, I like all sports. And I also like going to the cinema.

KEITH My name's Keith and I live in Hong Kong. I like computer games and water sports, like swimming and water polo. I go to clubs most weekends with my friends, because I like dancing.

SUSIE My name's Susie and I'm from Melbourne in Australia. I like going to the beach and dancing, and I also like swimming. I read novels and go to the theatre in my free time.

Lesson 11 Listening and writing, activity 2

Q Sam, tell me about your typical day.

SAM Well, I usually get up at about midday.

Q Midday!

SAM Yes, midday.

Q But that's...

SAM Twelve o'clock, yes, that's right. Then I...

Q But what's your job?

SAM Oh, I'm a musician. But I haven't got much work at the moment.

Q OK, so what do you do after you get up?

SAM Well, I usually have breakfast.

Q So, after breakfast...?

SAM ... after breakfast, I always meet my friends and we often play football. And then we usually have some lunch.

Q And what do you do after lunch?

SAM Well, we sometimes go to a concert or we play some music.

Q And then what? Do you go to bed?

SAM Well, I usually have dinner in a restaurant at nine or ten in the evening if I'm hungry, and then...

Q ... you go to bed?

SAM ... I always go to a club or a party.

Q And when do you go home?

SAM Four or five o'clock in the morning.

Q Four or five o'clock in the morning? You do this every day?

SAM Well, not Sundays.

Q Not Sundays. What do you do on Sunday?

SAM I always stay at home and telephone my friends.

Q Every Sunday?

SAM Yes, most people are at home on Sunday.

Q I see. And this is a typical day?

SAM Yes, well, it's typical for me.

Lesson 13 Grammar, activity 2

JO What can you do?

PAT I can run 100 metres in 15 seconds and I can use a computer.

JO So can I.

PAT But I can't cook very well.

JO Nor can I. But I can speak a foreign language.

PAT What language can you speak?

JO Spanish. Can you speak Spanish?

PAT No, I can't.

Lesson 13 Speaking and listening, activity 3

FRANK Hey Ann, look I've got that, er, I've got that list. Listen, I'll read some of them out. You can see the Great Wall of China from space.

ANN Yes, I think you can.

FRANK Yeah, so that's true.

ANN Yes.

FRANK Yes, that's true. Let's have a look at this one. Here's one. Cats can't swim.

ANN Erm... I think they can, actually.

FRANK Yes, so therefore, it's, that's false.

ANN Cats can't swim, that's, no, false.

FRANK That statement's false. Where's another... oh look here... chickens can't fly.

ANN Erm...

FRANK Can they fly?

ANN I'm not sure.

FRANK I don't think they can.

ANN No, they can't.

FRANK So, that's true?

ANN Yes.

FRANK Right, OK.

ANN Chickens can't fly.

FRANK Chickens cannot fly. Computers can write novels.

ANN I don't think they can.

FRANK No, I don't think they can.

ANN That's ridiculous.

FRANK So, what that's... er, we'll put there, we'll put no.

ANN No, false.

FRANK That's false. Cameras can't lie.

ANN Yes, they can.

FRANK Yes, so that's erm, that's false. That's false too. Erm... England can win the next World Cup competition.

ANN Oh, I don't know.

FRANK I think that's false. Do you think?

ANN I don't know. I... they could, couldn't they?

FRANK Well, they could win it.

ANN Yes. So, yes true.

FRANK Alright, we'll make it true. Right, we'll do that. Thin people can't swim very well.

ANN That's rubbish.

FRANK That's rubbish, they, er... they swim very well, so that's false. Erm... here's, here's a good one. You can never read a doctor's handwriting.

ANN Well, it's a bit of a cliché that, isn't it?

FRANK Yeah, it is.

ANN But I think it's false.

FRANK I think it's false, too. No. Erm, here we go. Oh, here's a good one. You can clean coins with Coke.

ANN I believe you can.

FRANK Yes, I've heard that too.

ANN I've never done it.

FRANK So, erm... yeah that's true.

ANN Let's say that's true.

FRANK OK, and the last one I've got here is, cats can see in the dark. I think that's erm...

ANN I think that's true.

FRANK I think it's true too. Yes, that's true.

ANN OK, let's say true.

Lesson 14 Vocabulary and listening, activity 3

Conversation 1

WOMAN Er, can you help me?

MAN Yes, of course.

WOMAN Where is the bank?

MAN There's a bank in Valley Road.

WOMAN How do I get there?

MAN Go along Prince Street, turn left up George Street. Turn left into Valley Road and the bank is opposite the cinema.

WOMAN Thank you.

Conversation 2

WOMAN Excuse me, where's the baker's?

MAN There's a baker's next to the cinema.

WOMAN How do I get there?

MAN Go up East Street and turn right into Valley Road. The baker's is on your left, opposite the florist.

WOMAN Thank you.

Conversation 3

WOMAN Can I help you?
MAN Er, yes. Where can I buy some aspirin?
WOMAN There's a chemist in Prince Street.
MAN Oh, how do I get to Prince Street?
WOMAN Go down George Street and turn right into Prince Street. The chemist is opposite the restaurant.
MAN Thank you.

Conversation 4

MAN Excuse me, where can I buy a newspaper?
WOMAN There's a newsagent in East Street.
MAN How do I get to East Street?
WOMAN Go along Prince Street. Turn right into East Street the newsagent is on your left.
MAN Thank you.

Lesson 15 Listening, activity 2

CAPTAIN Ladies and gentlemen, this is your captain speaking. I hope you're enjoying your flight to Rome this morning. At the moment, we're flying over the beautiful city of Zurich, in the centre of Switzerland. If you're sitting on the left-hand side of the plane, you can see the city from the window. We're flying at 12,000 metres and we're flying at a speed of 750 km/h. I'm afraid the weather in Rome this morning is not very good. It's snowing and there's a light wind blowing. Enjoy the rest of your flight. Thank you for travelling with us today.

Lesson 15 Grammar, activity 2

Conversation 1

MAN Everything all right, darling?
WOMAN Yes, everything's fine. The food is delicious.
MAN Good, I'm delighted. I'm very fond of this restaurant. The chef is excellent.
WOMAN Do you come here often?
MAN Ah, well, only with... with very special people.
WOMAN Such as?
MAN Well, such as business clients...

Conversation 2

WOMAN Look at those oranges! Let's have some, shall we?
MAN 1 But we've got a lot of fruit – bananas, apples, melons. We don't need oranges as well.
WOMAN And we can get some potatoes and eggs here. How much are the potatoes?
MAN 2 Thirty pence a kilo.
WOMAN OK, I'd like two kilos.
MAN 1 Two kilos of potatoes.
MAN 2 Two kilos of potatoes, there you go.

Conversation 3

MAN What's happened?
WOMAN I don't know.
MAN Excuse me, what's happening?
DRIVER I don't know.
MAN Well, how long are we staying here?
DRIVER I don't know.
MAN But I'm going to work. I'm late already. What am I going to do?
DRIVER I don't know.
MAN Do you know why you're beginning to annoy me?
DRIVER No.
MAN You keep saying 'I don't know.'

Progress check 11 to 15 Vocabulary, activity 5

Conversation 1

MAN Excuse me!
WOMAN Yes, sir. Can I help you?
MAN Yes, I'm looking for the men's department.
WOMAN It's over there!
MAN Thank you.

Conversation 2

MAN Can you carry this for me, please?
WOMAN Pardon? I didn't hear what you said.
MAN I said can you carry this for me, please?
WOMAN Yes, of course.

Conversation 3

MAN Oh, sorry!
WOMAN That's OK.

Lesson 16 Speaking and listening, activity 2

SPEAKER 1 Her name was Mrs Smith. She was a tall lady, and was very old, or so it seemed. But she was a very good teacher, and she was very kind to us.
SPEAKER 2 All my friends were at my party. I remember it was a sunny day, in June, and we were all in the garden, playing together. There was lots to eat and drink and lots of games to play. Then they all sang Happy Birthday.
SPEAKER 3 There was a large map of the world on the walls of the classroom, on the one side of the board, and on the other, there was a chart to show us how to write the alphabet.
SPEAKER 4 His name was Jack, and we were very good friends. Our mothers were very good friends too, so I saw Jack nearly every day for the first three or four years of my life. Then we were at different schools, so we weren't together so often.

Lesson 17 Vocabulary and listening, activity 4

JEAN OK, what do we need?
TONY We need some fruit and vegetables.
JEAN How about some oranges?
TONY OK, and we'll have some bananas.
JEAN Yes, there aren't any bananas. And let's get some apples.
TONY OK, apples. And we haven't got any onions.
JEAN A kilo of onions. That's enough. And some carrots?
TONY That's right, we haven't got any carrots. And let's get some meat.
JEAN Yes, OK. You like chicken, don't you?
TONY Yes, chicken's great. And we need some tomatoes.
JEAN OK, two kilos of tomatoes. Anything else?
TONY No. Oh, have we got any water?
JEAN No, we need a couple of litres of water and let's get some juice. That's it.

Lesson 17 Listening and speaking, activity 2

Q Is it true, Lisa, that you always have bacon and eggs for breakfast?
LISA Well, it used to be true, but it isn't true any more. People often have toast and cereal, jam, yoghurt, things like that, but not many have time to cook bacon and eggs. It's only in hotels when you get bacon and eggs – what we call a cooked breakfast, or an English breakfast.
Q Do you always have meat and vegetables at the main meal?
LISA Not always, no. In any case, I don't eat meat.
Q Do people in Britain drink a lot of wine?
LISA Yes, they do, but not at every meal. They don't drink wine at lunch and dinner every day. Perhaps we'll have a glass of wine at the weekend.
Q What do people drink then?
LISA Water, juice. Some families drink tea with their meals.
Q Do you drink *much* tea?
LISA Yes, people drink tea during the day. Many people will have five or six cups of tea a day. We usually have it with milk. Other people prefer coffee.
Q What's the main vegetable that you find at most meals?
LISA I suppose we often eat potatoes with our main meal. Potatoes are very popular. But we also eat a lot of pasta, but we don't eat pasta *and* potatoes!
Q Are there many people who don't eat meat in Britain?
LISA Yes, there are many vegetarians like me in Britain. Every restaurant will always offer several vegetarian dishes, and when people come to dinner you always check to see if there are any vegetarians.

Lesson 18 Listening and speaking, activity 1

MAN Whitney Houston was born in New Jersey in 1963. She started singing when she was eleven years old. As a teenager she worked with rhythm and blues singers such as Chaka Khan and Lou Rawls. In 1985 she had a hit with her first album, *Whitney Houston*. She received a Grammy Award with the song *Saving all my love for you*. In 1986 she was the first pop singer to sell ten million copies of her first two albums. In 1992 she appeared with Kevin Costner in the film *The Bodyguard*. The following year she had another hit with *I will always love you* from the film. She isn't married.

Lesson 19 Functions, activity 1

Conversation 1

MAN I've heard so much about your mother. Tell me about her. What's she like?

WOMAN Well, she's quite elderly now. She's medium-height, with short hair. She's still quite attractive, I think. What else can I say? Oh, well, you can see for yourself in an hour.

MAN I am looking forward to meeting her.

Conversation 2

WOMAN 1 So you've got a new boyfriend, I hear.

WOMAN 2 That's right.

WOMAN 1 What's he like, then.

WOMAN 2 Well, he's middle-aged and rather good-looking.

WOMAN 1 What does he look like?

WOMAN 2 Oh, he's quite tall, with dark hair.

WOMAN 1 And what does he do?

WOMAN 2 He's an accountant.

WOMAN 1 Oh.

Conversation 3

MAN And you say he's going to arrive by plane from Frankfurt.

WOMAN Yes, that's right.

MAN How will I recognise him? What's he like?

WOMAN Well, he's short and he's got dark curly hair. He's quite well-built.

MAN Anything else?

WOMAN Oh, and he's got a moustache.

MAN I see. How old is he?

WOMAN He's twenty.

MAN I suppose I should write his name on a card and hold it up as the passengers come out of the arrivals lounge.

WOMAN Oh, I'm sure you'll recognise him.

Conversation 4

MAN So, what's the new woman at work like?

WOMAN Oh, she seems very nice. She's quite young. About thirty. She's tall and slim with long fair hair. And she wears glasses.

MAN She sounds very nice. When do I meet her?

WOMAN Oh, she's not your type, John.

MAN Not my type? Tall? Slim? Long fair hair? Not my type?

WOMAN No, John.

MAN Why not?

WOMAN Well, have a look in the mirror, John.

MAN Some people think I'm very good-looking.

Lesson 19 Listening and speaking, activity 2

Q Would you say that the Irish are tall people, Kevin?

KEVIN No, they're not very tall, I suppose we think someone is quite tall when they're over one metre eighty, something like that.

Q And when does old age begin, in your opinion? How old are people when they're old?

KEVIN I suppose about sixty or seventy. No, it used to be sixty, but it's changing. No, I'd say seventy.

Q And middle age? When are people middle-aged?

KEVIN I'd say you're middle-aged when you're over forty.

Q And would you say that people are well-built or slim?

KEVIN Oh, it's hard to say. I suppose some of us are well-built, some of us are slim. Yes, I think it is difficult to say.

Lesson 20 Vocabulary and listening, activity 3

MAN Hi, Mary, Hi, Bill!

BILL Hello, how are you?

MAN Fine thanks, how was your trip?

MARY It was great. We had a wonderful time.

MAN Where did you go?

MARY Well, we flew to Paris, where we did everything the tourists usually do. We walked a lot by the river...

BILL And the weather was sunny...

MAN Great!

MARY And we visited the museums and galleries, and the Eiffel Tower...

BILL And did some shopping.

MARY Yes, we did some shopping.

BILL No, *you* did some shopping.

MARY OK, OK, I did a lot of shopping.

BILL Then we went to Venice, where we met some New Yorkers, and we did a lot of sightseeing with them.

MARY We had a nice time together.

BILL Then we went to Vienna...

MARY I've always wanted to go to Vienna.

BILL And we found a cheap hotel and what did we do there?

MARY We went to the Opera. It was wonderful. So romantic. And then we went to Budapest.

BILL Yeah, and Mary bought some souvenirs, while I had a steam bath. It was great.

MARY Except you lost your wallet.

BILL Yes, I lost my wallet, but someone found it and luckily it had the name of my hotel in it and I got it back.

MARY Isn't that great?

MAN Amazing.

MARY And then we flew back to Paris, and then flew home.

MAN Well, welcome home!

BILL & MARY Thank you.

Progress check 16 to 20 Sounds, activity 4

WOMAN Was Jonas born in Vienna?

MAN Was there anything to eat at the party?

WOMAN Did you see the football match last night?

MAN Were there lots of people on the bus?

WOMAN Did they arrive on time?

MAN Did she get home safely?

Lesson 21 Sounds

What did Agatha Christie write?

When was she born?

Where did she live?

Who did she marry?

When did she disappear?

Where did her husband find her?

When did he find her?

What did Sir Max Mallowan do?

Lesson 22 Listening and speaking, activity 2

SPEAKER 1 Oh, it was five months ago, in August, we were talking about it yesterday evening. It was a typical wedding day, which started with the excitement of getting dressed in my wedding dress. Then we drove to the church, and we arrived about five minutes late, which is the custom. Then the service started and after the service, there was a photographer. And finally we went to the wedding reception, which was wonderful, and we danced until about three in the morning. It was a wonderful day.

SPEAKER 2 It was last Thursday when I heard my exam results. After three years of hard work at university, at the end of the year, the university year, that is, finally the day arrived when I found out if I had passed the exam. I woke up early and went down to the examination building where they put up the exam results on lists. The building was closed so I waited about ten minutes with a number of other students. Then they opened the doors and we went in, and I found my name on the list. I'd passed! What a great feeling! I left the building, dancing and singing, people must have thought I was mad.

SPEAKER 3 It was in 1987, on the eleventh of December, that my parents had their golden wedding anniversary, that's fifty years of marriage. And we had a big party for them, all my brothers and sisters, all the grandchildren, nephews, nieces, friends and neighbours, there were about sixty people in all. We worked all day from nine to five preparing the food and decorating the house. And the funny thing was – they didn't know about the party until they got home. They walked in the door and everyone started singing. It was a great surprise!

Lesson 23 Vocabulary and listening, activity 5

MAN Hey, this is a great party.
WOMAN I'm glad you were able to come.
MAN You must introduce me to your friends. Which one is Harriet? I've heard so much about her.
WOMAN Well, can you see the woman by the door?
MAN Er, which one?
WOMAN The one in the jeans.
MAN No, which door?
WOMAN There is only one door.
MAN Oh, right! Yes, I see. The one in the jeans and the T-shirt.
WOMAN Yes, she's standing by the door and talking to the middle-aged man. She's smiling at him.
MAN Great smile. And what about John?
WOMAN John is sitting down in the armchair by the window.
MAN Is he the one wearing the sweater and trousers?
WOMAN No, John's wearing the shirt and tie.
MAN OK, I see him. Nice tie. Love that yellow.
WOMAN And Edward is the man over by the window. He's laughing at something. Probably laughing at John's tie.
MAN The man in the trainers with the blue shirt and black trousers?
WOMAN That's Edward!
MAN Right! And Louise?
WOMAN Louise is standing by the television. She's wearing a black dress.
MAN Yes, I can see her. Hey, she's smiling at me.
WOMAN She smiles at everyone. She wears glasses, but she hasn't got them on at the moment.
MAN No, I'm sure she's smiling at me...

Lesson 24 Reading and listening, activity 3

Speaker 1
MAN Well, we find that our jobs take up a lot of time, and when the weekend comes, we're very tired.
WOMAN But the trouble is, we don't have any social life now.
MAN So our New Year's Resolution is we're going to invite more friends for dinner.
WOMAN Yes, because we don't entertain much. And we'd like people to invite us as well.

Speaker 2
MAN I've got a year off before I go to university, and I don't know much about foreign countries, so I'm going to travel around Europe. I always stay in Britain for my holiday, and I'd like to see how our neighbours live.

Speaker 3
WOMAN I spend my life driving to school, teaching and then coming home. So I'm going to get fit because I don't take enough exercise. I'm going to start running in the evening when I get home.

Speaker 4
MAN We've got a family now, and we need more space for all the things that young children need. Our resolution is that we're going to move.
WOMAN Yes, because our house is too small for four people. The trouble is, we still like it here, though.
MAN Yes, it's going to be hard to leave this house.

Lesson 25 Vocabulary and listening, activity 5

ASSISTANT Good afternoon. Can I help you?
CUSTOMER Good afternoon. Yes, I'd like a burger with fries and a Coke, please.
ASSISTANT Would you like a regular or a large Coke?
CUSTOMER A regular, please.
ASSISTANT Would you like anything else?
CUSTOMER Yes, I'd like an ice cream, please.
ASSISTANT What flavour would you like?
CUSTOMER Strawberry, please.
ASSISTANT OK.
CUSTOMER How much is that?
ASSISTANT That's four pounds fifty.
CUSTOMER Here you are.
ASSISTANT Thank you.

Lesson 26 Speaking and listening, activity 4

ASSISTANT Can I help you?
CUSTOMER Yes, I'm looking for a sweater.
ASSISTANT We've got some sweaters over here. What colour are you looking for?
CUSTOMER This blue one is nice.
ASSISTANT Yes, it is. Is it for yourself?
CUSTOMER Yes, it is. Can I try it on?
ASSISTANT Yes, go ahead.
CUSTOMER No, it's too small. It doesn't fit me. Have you got one in a bigger size?
ASSISTANT No, I'm afraid not. What about the red one?
CUSTOMER No, I don't like the colour. Red doesn't suit me. OK, I'll leave it. Thank you.
ASSISTANT Goodbye.

Lesson 26 Vocabulary and listening, activity 4

Conversation 1
ASSISTANT Can I help you, sir?
CUSTOMER Yes, I'm looking for something for my wife. Have you got any perfume?
ASSISTANT Yes, sir. Try this. It's our new perfume for this season. It's called *Reward*.
CUSTOMER Can I try it? Mm. It's very good. How much is it?
ASSISTANT This bottle costs £37.
CUSTOMER Hm. Have you got it in a smaller bottle?

Conversation 2
CUSTOMER Excuse me.
ASSISTANT Yes, madam, can I help you?
CUSTOMER I'd like a box of chocolates, for my friend.
ASSISTANT Yes, madam. 500 grams?
CUSTOMER How much is that?
ASSISTANT Ten pounds. These are particularly good chocolates.
CUSTOMER OK, I'll have them.

Lesson 27 Listening and speaking, activity 2

OFFICIAL Good afternoon, how can I help you?
CUSTOMER I've lost my bag. I put it down somewhere in the market, and forgot about it.
OFFICIAL Well, don't worry, madam, it'll turn up. I'll just take down some details. Could you tell me your name, please?
CUSTOMER Yes, my name's Jill Fairfield.
OFFICIAL Jill Fairfield. Is that Mrs Fairfield?
CUSTOMER Ms Fairfield.
OFFICIAL OK, and your address?
CUSTOMER 32, Burn Road, Manchester.
OFFICIAL Burn Road, Manchester. OK, and your telephone number?
CUSTOMER 679 5453.
OFFICIAL Right and you say it was a bag that you lost?

CUSTOMER That's right.
OFFICIAL And you lost it in Chester market?
CUSTOMER That's correct. At about ten in the morning.
OFFICIAL And that was today, right?
CUSTOMER Yes.
OFFICIAL So, Thursday the twenty-first of July. And can you describe it to me?
CUSTOMER Well, it was large, square and it was made of black nylon.
OFFICIAL And was there anything in it?
CUSTOMER Yes, there was my purse, a calculator, an address book, a newspaper and a comb.
OFFICIAL Right, now, let me see. Was it this one?
CUSTOMER No, that's not mine.
OFFICIAL Is this yours?
CUSTOMER Yes, that's mine. Thank you very much.

Lesson 27 Listening and speaking, activity 5

Conversation 1
CUSTOMER Excuse me!
OFFICIAL Yes, sir, good morning.
CUSTOMER I'd like to report the loss of a coat.
OFFICIAL Yes sir. And your name is...?
CUSTOMER Ken Hamilton.
OFFICIAL Ken Hamilton, all right. And your address?
CUSTOMER 13, Dock Lane, London.
OFFICIAL And your phone number?
CUSTOMER 75859.
OFFICIAL And when did you lose it?
CUSTOMER On the thirty-first of May around four in the afternoon, I must have left it on the train.
OFFICIAL And what was it like?
CUSTOMER It was made of red leather.
OFFICIAL Red leather! Was it... a man's coat, sir?
CUSTOMER Er, well, yes... er...
OFFICIAL I see. Now, did you lose anything else?

Conversation 2
CUSTOMER And I must have lost it then.
OFFICIAL Just say your name again, madam.
CUSTOMER Mary Walter.
OFFICIAL And your address and phone number?
CUSTOMER 21, Tree Road, Leeds, 75889.
OFFICIAL And it was a black plastic bag, you say?
CUSTOMER Yes.
OFFICIAL And you last saw it on the twentieth of March at two in the afternoon?
CUSTOMER Yes, in the supermarket.
OFFICIAL And what was in it?
CUSTOMER All my shopping and my purse.

Conversation 3
CUSTOMER J-O-S-E-P-H.
OFFICIAL OK, Mr Joseph. And the address?
CUSTOMER Where I left it?
OFFICIAL No, your address.
CUSTOMER 33, James Street, Bath.
OFFICIAL Do you have a daytime telephone number?
CUSTOMER Yes, 56778.
OFFICIAL What was this box of cigars made of?
CUSTOMER Well, it was wooden and square.
OFFICIAL I see. And you left it on the bus?
CUSTOMER Yes, yesterday morning at about eleven o'clock.
OFFICIAL OK, about eleven o'clock on the seventeenth of October.

Lesson 28 Vocabulary and listening, activity 2

Conversation 1
DOCTOR Good morning. How are you?
PATIENT Fine, thanks.
DOCTOR So, if you're fine, why are you here to see me?
PATIENT No, what I meant was, oh, it doesn't matter. I've got a headache. I seem to have it all the time.
DOCTOR I see. Any other symptoms?
PATIENT Well, I've got a cough as well.
DOCTOR Do you smoke?
PATIENT Yes I do. And I feel tired all the time.
DOCTOR OK, let's have a look.

Conversation 2
DOCTOR And what seems to be the matter with you?
PATIENT I feel sick and I've got a stomachache.
DOCTOR Let me see. Have you got a headache?
PATIENT Yes, I have.
DOCTOR You look rather hot. Yes, you've got a bit of a temperature. I think it must be something you ate yesterday.
PATIENT I only had a sandwich yesterday.
DOCTOR What kind of sandwich?
PATIENT It was a cheese sandwich.
DOCTOR Well, it's probably nothing serious, but I'll give you some medicine...

Conversation 3
DOCTOR And what seems to be the trouble?
PATIENT I've hurt my leg.
DOCTOR How did you do that?
PATIENT In a game of football.
DOCTOR Football! Don't you think you're too old to play football?
PATIENT Well, I'm only seventy-three.
DOCTOR Really! Well, let me see now...

Lesson 29 Reading and listening, activity 2

Q Karl, tell me something about Sweden. Is it a large country?
KARL Yes, it is quite large, at least for a country in Europe. It's almost 450,000 square kilometres.
Q 450,000. Yes, that's quite big.
KARL Yes, it's nearly twice the size of Britain.
Q I see. And what's the coldest month? January, I suppose?
KARL Yes, January is quite cold. The temperature is about minus three degrees Celsius.
Q And how hot is it in the summer?
KARL Actually, it gets quite warm. In July, the average temperature is eighteen degrees.
Q And what's the average rainfall?
KARL Well, for Sweden, it's 535mm, but it's less in Stockholm.
Q And what's the population?
KARL There are over eight million people. Not so many for such a large country.
Q And what about the armed services? How many troops are there?
KARL We only have about 65,000 soldiers.
Q 65,000, I see. And when do children start school?
KARL They start at the age of seven and continue for ten years, until they're seventeen.
Q Ten years.
KARL Of course, some students go on to university.

Lesson 30 Vocabulary and listening, activity 5

KATY What's your favourite sport, Andrew?
ANDREW Well, I like most sports, but I suppose I like football most of all. Like most people.
KATY Yes, I suppose football is the most popular sport. Personally, I don't like football. I don't enjoy competitive sports. I like cycling and horseriding.
ANDREW Isn't horseriding very expensive?

KATY Yes, it's more expensive than cycling.

ANDREW I think horseriding is the most expensive sport. What do you think is the most tiring sport?

KATY Well, horseriding is very tiring.

ANDREW Do you think it's more tiring than, say, tennis?

KATY Oh, yes, I'm exhausted after I've been horseriding. What about you?

ANDREW Well, for me tennis is the most tiring. What do you think is the most dangerous sport?

KATY I think hanggliding is very dangerous.

ANDREW Well, that's what many people think. But you know, there are more accidents to do with windsurfing than there are with hanggliding.

KATY I didn't know that. Which is the most difficult sport, in your opinion?

ANDREW How about climbing? I think climbing is very hard.

KATY Well, I think skiing is more difficult than climbing.

ANDREW No, I don't agree. Climbing looks incredibly difficult.

KATY And what do you think is the most exciting sport?

ANDREW Well, tennis, I think. What about you?

KATY It has to be motor racing. Motor racing is the most exciting sport for me.

Lesson 31 Listening, activity 1

JAMES Well, in Australia, you needn't ask if you want to take a photograph of someone you don't know. Even someone you do know will not usually mind if you take a photo.

Yes, I've heard about this custom with shoes in Japan. Actually, it's quite a good idea. But you needn't take your shoes off when you go into someone's house in Australia, unless of course, they're very dirty. I usually change my shoes when I get home, but I don't take them off when I visit people.

And women needn't cover their heads in Australia, most of the time. Sometimes, in certain churches, women wear a hat or a scarf on their heads, but they needn't do so in the street or at work.

Pointing at people, yes, that's true in Australia, you mustn't point at people. If you do, people think you're rather rude. It's the same in Britain, I think.

In Australia you can look people in the eye, though. It shows you're interested in them and what they're saying. It's a sign of politeness. But you mustn't look people in the eye for long, as they begin to feel uncomfortable.

Well, you *can* kiss in public in Australia, but not many people do so. We're fairly relaxed in my country, so almost anything is OK, but, well... let's say it's only young people who kiss in public – it's not forbidden, but it's not very common.

If you give a gift, you needn't use both hands. You can use just one hand. There aren't any rules about this sort of thing.

And, no you needn't shake hands with everyone when you meet them in Australia. You can shake hands when you meet someone for the first time, in fact, it's bad manners if you don't, but not every time, no.

Lesson 31 Sounds, activity 1

Children mustn't play near the road.
You must be quiet in a library.
You must keep your wallet in a safe place.
Men must take off their hats in a church.
You mustn't give a gift with one hand in Taiwan.
You mustn't wear shoes in a Japanese home.

Lesson 33 Listening and vocabulary, activity 3

ALAN Hi! How's your day been?

BARRY Awful, absolutely awful.

ALAN I'm sorry to hear that. What's happened?

BARRY Well, I've hurt my back.

ALAN Your back! How did you hurt it?

BARRY I tried to lift a box of books.

ALAN A box of books! I'm not surprised you hurt yourself trying to lift a

box of books. Have you been to the doctor yet?

BARRY No, not yet.

ALAN Well, I think you should go immediately. And what else has happened?

BARRY I've lost my wallet.

ALAN Your wallet? Where did you lose it?

BARRY At the bus stop, I think.

ALAN Have you been back to the bus stop yet?

BARRY No, I haven't.

ALAN And have you heard your exam result?

BARRY Yes, I have.

ALAN Have you passed?

BARRY No, I've failed it.

ALAN Oh dear, it's been one of those days for you, hasn't it?

Lesson 33 Listening and vocabulary, activity 6

Conversation 1

WOMAN Have you seen my bag?

MAN No, where did you leave it?

WOMAN Right here under my desk. But it's not there now.

MAN When did you last see it?

WOMAN This morning.

MAN Oh no! Are you thinking what I'm thinking.

WOMAN I think someone has stolen it.

MAN That's right. I think someone has stolen your bag.

Conversation 2

MAN Come on!

WOMAN I can't run any faster.

MAN It's not far. We can make it.

GUARD Sorry, sir, it's too late.

MAN What do you mean... too late?

WOMAN We've got to catch the Waterloo train.

GUARD I'm sorry but you've just missed the Waterloo train. It's just left.

WOMAN Oh no! Now what are we going to do?

Conversation 3

MAN And Williams comes along the right wing and passes to Franks and Franks runs with the ball, round Gray, and flips it over the goalkeeper's head and he's scored, he's scored a goal for United. What a brilliant goal!

Conversation 4

WOMAN Aargh!

MAN What's the matter! What have you done!

WOMAN I've cut my finger.

MAN You've cut your finger! Oh, I am sorry. Don't worry. I know exactly what to do.

WOMAN What?

MAN I should put my head between my knees so I don't faint at the sight of blood.

WOMAN Thanks. Thanks very much for your help.

Lesson 35 Vocabulary and sounds, activity 2

Conversation 1

MAN Are you asleep?

WOMAN No, not yet. But I'm very tired.

MAN Did you hear that?

WOMAN Hear what?

MAN That noise.

WOMAN No, I didn't.

MAN I think I'll find out what it is.

Conversation 2

MAN Could you move your car, please?

MAN 2 What?

MAN I said, could you move your car?

MAN 2 Why should I?

MAN Because it's in my way.

Conversation 3

WOMAN I'm so sorry to trouble you, but could you tell me how to get to Thames Street?

MAN Thames Street. Well, if I were you, I wouldn't start from here.

Conversation 4

TEACHER Quiet! ... I said quiet! Stop shouting, Smith! Sit down, White! Get your books out! Will you all be quiet!

Lesson 35 Listening and speaking, activity 2

JENNY Look, there are these questions here about how you got on at school. Shall we, shall we just go through them?

GAVIN Yes, let's.

JENNY OK, so, did you always work very hard?

GAVIN Er, well I certainly worked pretty hard at the subjects I enjoyed, yes I did. What about you?

JENNY Yes, I did actually, I think I worked very hard, yeah. And what about this other one?

GAVIN Did, yeah, did you always listen carefully to your teachers?

JENNY Erm... no I don't think I did. No, I think I was quite disruptive, actually. What about you?

GAVIN Well, I think I did listen to the teachers certainly when I got to the level where I was doing the subjects that I enjoyed.

JENNY Yeah, OK, this next one says, did you always behave well?

GAVIN I don't think I did always behave that well. I was erm, a bit, er, a bit of a tearaway.

JENNY Mm. Well, I think I was pretty well-behaved on the whole, so I'd say yes, yeah.

GAVIN Good for you! Did you pass your exams easily?

JENNY Erm, no I can't say I did, no, I, I found them quite a struggle, actually. What about you?

GAVIN I didn't pass them that easily, because I worked hard but also I found it very difficult writing all that amount of material in such a short amount of time.

JENNY Yeah, yeah, exactly. What about this one, then? Did you always write slowly and carefully?

GAVIN Quite slowly. Essays took a long time to write and I suppose I took quite a bit of care, yes.

JENNY Yes, I agree. I was also, I was very careful and erm, yeah, yeah I was quite methodical.

GAVIN And did you think your school days were the best days of your life?

JENNY Um, no, no I can't say they were. What about you?

GAVIN No, I went away to boarding school when I was quite young and I didn't like that. No, I, they weren't the best days of my life.

Lesson 36 Listening, activity 2

OFFICER Can I help you?

CUSTOMER Yes, I'd like a ticket to Birmingham.

OFFICER When do you want to travel? It's cheaper after 9.30.

CUSTOMER I'll travel after 9.30.

OFFICER Single or return?

CUSTOMER I'll have a single ticket, please.

OFFICER That'll be thirty pounds exactly. How would you like to pay?

CUSTOMER Do you accept credit cards?

OFFICER I'm afraid not.

CUSTOMER Well, I'll pay cash, then. Will there be refreshments on the train.

OFFICER Yes, there will.

CUSTOMER Can I have a ticket for the car park as well?

OFFICER That'll be thirty-two pounds in all.

CUSTOMER Thank you.

Lesson 37 Vocabulary and listening, activity 4

FORECASTER And here's the weather forecast for the rest of the world. Athens, cloudy twelve degrees. Bangkok, cloudy thirty degrees. Cairo sunny sixteen degrees. Geneva, cloudy ten degrees. Hong Kong, cloudy twenty degrees. Istanbul, rainy seven degrees. Kuala Lumpur, sunny thirty-five degrees. Lisbon, cloudy eleven degrees. Madrid, rainy seven degrees. Moscow, snowy minus ten degrees. New York, sunny zero degrees. Paris, snowy minus six. Prague, sunny minus two. Rio, cloudy minus 29. Rome, rainy nine degrees. Tokyo, snowy minus four degrees and finally Warsaw, cloudy minus eight degrees.

Lesson 37 Reading and listening, activity 4

Q And we have a scientist with us today to discuss various predictions about our weather in the future. Professor Stein, what do you think about the latest report about our climate?

STEIN I think the predictions are, in general, very accurate.

Q So you think temperatures will rise in the future?

STEIN Yes, in the next twenty-five years, they'll rise by two to six degrees.

Q And what will the consequences of that be?

STEIN Well, the ice at the North and South poles will melt and the sea level will rise.

Q Do you think whole countries will disappear?

STEIN No, I don't. That won't happen for another hundred or more years.

Q And will there be enough fresh water for everyone?

STEIN Yes, there will. But it won't come from rainfall, which will decrease in general, but from the sea with the salt taken out.

Q Will fresh water cost more?

STEIN Yes, certainly. And this will mean that factory goods will cost more to produce.

Q What effect will this have on the economy? Will it get worse?

STEIN No, I don't think so. But we'll have to change our lifestyles in the future.

Lesson 38 Listening, activity 1

FRANK So, Sally, this quiz then, shall we have a go? Right, number one, coffee is grown in a, Brazil, b, England, c, Sweden. What do you think?

SALLY Oh, well, that's got to be Brazil hasn't it?

FRANK I think so, yeah, that's, a.

SALLY OK, number two, Daewoo cars are made in Switzerland, Thailand or Korea?

FRANK Korea, definitely.

SALLY Yeah? OK, so that's, c.

FRANK Number three, Sony computers are made in Japan, USA or Germany?

SALLY Japan.

FRANK Mm... a, then.

SALLY OK, erm, number four. Tea is grown in a, India, b, France or c, Canada?

FRANK I think it must be India, mustn't it. Don't you think?

SALLY Yeah, definitely, so that's, a for that one.

FRANK OK, number five, tobacco, where's tobacco grown then, Norway, Iceland or the USA? Well it's too cold for Iceland.

SALLY Yeah, it's the USA.

FRANK The USA, that's right, c then.

SALLY OK, Benetton clothes are made in Italy, France or Malaysia?

FRANK Benetton, I think, that's Italy, don't you?

SALLY Yeah, I think it is. Yeah.

FRANK a.

SALLY a. ... er, number seven, Roquefort cheese is made in a, Germany, b, Thailand or c, France?

FRANK Erm.

SALLY It's not Germany, is it?

FRANK No, I don't think it's Germany, I think Roquefort is France.

SALLY Yeah, France, OK.

FRANK Right, number eight. The atom bomb was invented by the Japanese, the Americans or the Chinese?

SALLY The Americans.

FRANK Yeah, the Americans, b.

SALLY Er, *Guernica* was painted by Picasso, Turner or Monet?

FRANK Um, that's Picasso, I think, that one.

SALLY Is it?

FRANK Yeah, yeah, famous painting that, a.

SALLY Right, the West Indies were discovered by Scott of the Antarctic,

Christopher Columbus or Marco Polo?

FRANK I think that's Christopher Columbus, was the one.
SALLY Was it, are you sure?
FRANK Yeah, I think, pretty sure.
SALLY OK, that's b.
FRANK OK, telephone, who invented the telephone, Bell, Marconi or Baird? Baird invented the television, I think.
SALLY Oh, it's Bell.
FRANK Bell?
SALLY Yeah, definitely.
FRANK OK, twelve, *Romeo and Juliet*, who wrote *Romeo and Juliet*? That's simple, isn't it?
SALLY Yeah.
FRANK Go on, then.
SALLY Well, it wasn't Ibsen.
FRANK No, it's got to be Shakespeare.
SALLY Yeah, and it wasn't Primo Levi.
FRANK No, that's, b, OK.
SALLY Number thirteen. The Blue Mosque in Istanbul was built by a, Sultan Ahmet I, b, Ataturk or c, Suleyman the Magnificent?
FRANK Well, I know this one.
SALLY Do you?
FRANK Yup, it was Ataturk.
SALLY No, Sultan Ahmet I.
FRANK Oh. *Yesterday*, right who composed *Yesterday*, Paul McCartney, John Lennon, well, it certainly wasn't Mick Jagger.
SALLY No.
FRANK So Paul McCartney or John Lennon? I think I know.
SALLY I think it's, a, Paul McCartney
FRANK Mm. And finally, the pyramids, who were they built by, the Pharaohs, the sultans or the council? Wasn't my local council!
SALLY No, erm...
FRANK I think it was, it's...
SALLY It's the Pharaohs isn't it? Yeah, that's, a.

Lesson 39   Listening and reading, activity 2

CHRIS Good afternoon.
RECEPTIONIST Good afternoon. Can I help you?
CHRIS Have you got any beds for tonight?
RECEPTIONIST Yes, I think so. Sorry, but I've just started work at the hostel. How long would you like to stay?
CHRIS We'll stay for just one night.
RECEPTIONIST Yes, that's OK.
TONY Great!
RECEPTIONIST How old are you?
TONY We're both sixteen.
RECEPTIONIST One night's stay costs £6.50 each.
CHRIS Is it far from the hostel to the centre of Canbury.
RECEPTIONIST Yes, it's two kilometres. It takes an hour on foot.
TONY Is there a bus service?
RECEPTIONIST I think so. It takes about fifteen minutes. There's a bus every hour.
TONY When does the last bus leave the city centre?
RECEPTIONIST I think it leaves at nine o'clock in the evening. There's not much to do in the evening.
CHRIS We're very tired. We need an early night. What time does the hostel close in the morning?
RECEPTIONIST Er, at eleven am. Where are you walking to?
CHRIS We're going to Oxton. Are you serving dinner tonight?
RECEPTIONIST Yes, we're serving dinner until eight o'clock. And breakfast starts at seven-thirty.
TONY And where's the next hostel?
RECEPTIONIST I'm not sure. I think it's Kingscombe, which is about ten kilometres away. I started work last Monday so I'm very new here.

Lesson 39   Listening and reading, activity 4

CHRIS It's very strange. She said one night's stay cost £6.50, but it costs £6.15.
TONY Yes, and she said it was two kilometres to the city centre.
CHRIS But, in fact, it's three kilometres.
TONY And she said the last bus left at nine o'clock. But it leaves at eight o'clock.

Lesson 39   Grammar, activity 3

CHRIS And she said the bus took fifteen minutes. But in fact, it takes ten minutes.
TONY And she said the hostel closed at eleven am, but it's open all day.
CHRIS It seems that they serve dinner from six to seven.
TONY But she said they were serving until eight o'clock. And she also said breakfast started at seven-thirty...
CHRIS ... when, in fact, it says here that breakfast starts at seven.
TONY And she said that Kingscombe was the next hostel, but it isn't. It's Charlestown.
CHRIS And finally she said that Kingscombe was ten kilometres away. But it's fifteen kilometres.

Lesson 40   Reading and listening, activity 6
Ruth's Parents

The next afternoon, Jan went to Ruth's house for tea.
'How do you do, Mr and Mrs Clark,' Jan said.
'Sit down, Jan,' said Mrs Clark. 'Would you like a cup of tea?'
'Yes, please,' said Jan. He didn't feel very comfortable.
'With sugar?' said Mrs Clark.
'Yes, please,' Jan answered. 'Sugar, but no milk.'
'No milk?' said Mrs Clark. 'But everybody drinks milk in tea!'
'Oh...' said Jan. 'Well, in Poland, we never have milk in our tea.'
Mr and Mrs Clark looked at Jan. 'These foreigners have strange ideas,' said Mr Clark.
Jan stayed for about an hour. Mr Clark spoke English very quickly and Jan did not always understand.
Outside the door, Jan said to Ruth, 'Your parents don't like me very much.'
'Don't be silly, Jan,' said Ruth. 'My parents haven't met many foreigners. It's all right.'
'OK, Ruth,' said Jan. 'I'll see you tomorrow.' And he walked away. But he felt unhappy.
Later that evening Ruth asked, 'Well, Mum, did you like Jan?'
Ruth's mother said, 'Well, he didn't speak English very well. Your father and I liked Bill. What's wrong with an English boyfriend? And Jan is going back to Poland soon.'
'But I don't like Bill any more,' shouted Ruth and ran out of the room. 'I don't like Bill,' she said to herself, 'but I do like Jan. Maybe I love him.'

Macmillan Education
Between Towns Road, Oxford OX4 3PP

Companies and representatives throughout the world.

ISBN 0 435 24205 9

Text © Simon Greenall, 1997.
First published 1997.
Design and illustration © Macmillan Publishers Ltd, 1997.

Designed by Stafford & Stafford
Cover design by Stafford & Stafford
Cover illustration by Martin Sanders

Illustrations by:
Adrian Barclay (Beehive Illustration), pp27, 78
Paul Beebee (Beehive Illustration), pp36, 37
Gillian Hunt (Beehive Illustration), pp4, 38, 92, 100, 101
Sarah McDonald, pp94, 103
Martin Sanders, pp11, 13, 20, 25, 27, 29, 32, 34/35, 37, 46, 47, 49, 58, 60, 64, 73, 76, 80, 84, 87, 88, 90, 91, 97, 98, 99, 101
Simon Smith, pp31, 59, 61, 64, 85, 99
Stafford & Stafford pp22, 23

Commissioned photography by:
Chris Honeywell pp10/11, 14, 40/41, 44, 45, 50, 54, 62, 94, 102
p5 ©HELT; p16 ©HELT(x3); M.Van Gelderen p100

Author's Acknowledgements

I am very grateful to all the people who have contributed towards *Reward* Elementary. Thank you so much to:
– All the teachers I have worked with on seminars around the world, and the various people who have influenced my work.
– James Richardson for the happy and efficient work he has done on producing the tapes, and the actors for their voices.
– Philip Kerr, for his comments on the material, which are especially helpful and well-considered.
– Sue Side, Maria Zeny, Sue Bailey, Sue Watson, Marcella Banchetti, Taska Eszter, Steve Bilsborough, Etienne Andre, Brian Waine, Burt Johnson, Miguel Almarza and Jim Scrivener for their reports on the material. I have tried to respond to all their suggestions, and if I have not always been successful, then the fault is mine alone.
– Simon Stafford for his usual, skilful design.
– Douglas Williamson for his efficient design management.
– Pippa McNee for tracking down some wonderful photos.
– Catherine Smith for her support and advice and her sensitive management of the project.
– Angela Reckitt for her attention to detail, her contribution to the effectiveness of the course, and her calm, relaxed style which makes work such a pleasure.
– and last, but by no means least, Jill, Jack and Alex.

Acknowledgements

The author and publisher would like to thank the following for their kind permission to reproduce material in this book: Heinemann Educational, a division of Reed Educational and Professsional Publishing Limited, for an extract from *Dear Ruth...Love Jan* by N. McIver; Reed Books Limited, for an extract from *The Return of Heroic Failures* by Steven Pile, published by Secker & Warburg Limited, l988; Times Publishing Group, Singapore, for an extract from *Culture Shock! USA* by Esther Wanning, published by Times Editions.

Photographs by: Anthony Blake Photo Library/Phototeque Culinaire pp80, Anthony Blake p105, Art Directors & Trip Photographic Library pp3, 20, 74, 103, British Rail International p57; Carlton TV p26; Corbis p86; European Passenger Services photograph p28; Eye Ubiquitous p74; Ronald Grant Collection p43; Chris Honeywell p33; Hulton Getty p50; Image Bank pp16, 30, 39, 70, 71, 76, 82, 88, 103; Images Colour Library pp7, 16, Lake School of English pp8, 9; Louis Psihoyos/Material World p18; Jeremy Hartly/Panos Pictures p15; Pictor International-London pp2, 103, 105, Rex Features p39, South American Pictures/Tony Morrison p2; Frank Spooner Collection p42; Stockphotos p30; Tony Stone Images pp2, 16, 19, 20, 30, 47, 56, 69, 70, 74, 82, 88, 89, 103, 105; Superstock p105; Universal Pictorial Press p39; Zefa Pictures p30

The publishers should also like to thank Celia Bingham, Nick Blinco, British Rail International, Carlton TV, Tim Cater, Aleeta Cliff, Tim Friers, Louis Harrison, Sue Kay at The Lake School of English, Helen Kidd, Sarah McDonald, Jamie McNee, Jason Mann, Andrew Oliver, Philip and Valerie Opher, Anthony Reckitt, Rebecca Smith, Martha and Rebecca Stafford, Julie Stone, Lydia Trapnell, Douglas Williamson, Chris Winter and Verley Woodley.

While every effort has been made to trace the owners of copyright material in this book, there have been some cases when the publishers have been unable to contact the owners. We should be grateful to hear from anyone who recognises their copyright material and who is unacknowledged. We shall be pleased to make the necessary amendments in future editions of the book.

Printed in Thailand

2008 2007 2006 2005 2004
25 24 23 22 21 20 19 18